EXTREME COSMOS

A Guided Tour of the Fastest,
Brightest, Hottest, Heaviest, Oldest,
and Most Amazing Aspects of
Our Universe

BRYAN GAENSLER, PhD

A PERIGEE BOOK

A PERIGEE BOOK
Published by the Penguin Group
Penguin Group (USA) Inc.
375 Hudson Street, New York, New York 10014, USA

Penguin Group (Canada), 90 Eglinton Avenue East, Suite 700, Toronto, Ontario M4P 2Y3, Canada (a division of Pearson Penguin Canada Inc.) • Penguin Books Ltd., 80 Strand, London WC2R 0RL, England • Penguin Group Ireland, 25 St. Stephen's Green, Dublin 2, Ireland (a division of Penguin Books Ltd.) • Penguin Group (Australia), 250 Camberwell Road, Camberwell, Victoria 3124, Australia (a division of Pearson Australia Group Pty. Ltd.) • Penguin Books India Pvt. Ltd., 11 Community Centre, Panchsheel Park, New Delhi—110 017, India • Penguin Group (NZ), 67 Apollo Drive, Rosedale, Auckland 0632, New Zealand (a division of Pearson New Zealand Ltd.) • Penguin Books (South Africa) (Pty.) Ltd., 24 Sturdee Avenue, Rosebank, Johannesburg 2196, South Africa

Penguin Books Ltd., Registered Offices: 80 Strand, London WC2R 0RL, England

While the author has made every effort to provide accurate telephone numbers, Internet addresses, and other contact information at the time of publication, neither the publisher nor the author assumes any responsibility for errors, or for changes that occur after publication. Further, the publisher does not have any control over and does not assume any responsibility for author or third-party websites or their content.

First American edition: July 2012
Originally published in Australia by University of New South Wales Press in 2011.

Perigee trade paperback ISBN: 978-0-399-53751-6

PRINTED IN THE UNITED STATES OF AMERICA

10 9 8 7 6 5 4 3 2 1

Most Perigee books are available at special quantity discounts for bulk purchases for sales promotions, premiums, fund-raising, or educational use. Special books, or book excerpts, can also be created to fit specific needs. For details, write: Special Markets, Penguin Group (USA) Inc., 375 Hudson Street, New York, New York 10014.

ALWAYS LEARNING   PEARSON

JUN 29 2012

To Fineas, for asking questions that I couldn't answer.

CONTENTS

PREFACE

When I first came up with the idea that eventually became this book, I thought it would be an easy thing to write. Most of the ideas and stories were already in my head, and it was just going to be a case of putting them down on paper. But the reality was far more complicated: The answers usually turned out to be far less simple than I first thought, and sometimes I realized that I didn't really understand the topic at all. The final product ended up taking two years of solid effort, and was only made possible thanks to input from a wide range of people and sources.

First and foremost, I thank the worldwide community of astronomers, whose passion and enthusiasm for their field have led to all the discoveries described here. I also must acknowledge the incredible resource that is the NASA Astrophysics Data System. This spectacular database contains an index of virtually every scientific article ever published in astronomy, and proved to be invaluable in tracking down and verifying the many results and calculations that I needed. I further thank the many astronomers who generously provided me with additional data and information: Matthew Bailes, Tim Bedding, Chris Blake, Warren Brown, Iver Cairns, Paul Crowther, Glennys Farrar, Lilia Ferrario, Craig Heinke, David Helfand, Rob Hollow, Michael Ireland, Melanie Johnston-Hollitt, Geraint Lewis, Charley Lineweaver, Erik Mamajek, Don Melrose, Michael Scholz, Peter Tuthill, Gentaro Watanabe, Mike Wheatland, and Matias Zaldarriaga.

I am grateful to Phillipa McGuinness, Jane McCredie, and their team at NewSouth Publishing for their role in bringing this book to fruition, and for their patience with a manuscript that was a long time in the making. I offer special thanks to Stephen Pincock, who sought me out, planted in my head the idea of writing a book, worked with me to develop the concept of *Extreme Cosmos*, and provided careful and thoughtful feedback on every chapter. I also thank Chris Hales, who enthusiastically sourced and researched many of the details, and who also offered many thoughtful comments on the text.

Finally, I offer my deepest, most sincere thanks to the amazing Laura Beth Bugg who, as always, has been my muse and inspiration. She took on the role of single parent while I sank so much time into this book, encouraged me to keep going when I was ready to give up, and pored over every word I wrote. She helped me to make this a work I can be proud of—I could not have done this without her.

Bryan Gaensler

INTRODUCTION

As a child, I loved reading about science. Descriptions of bizarre, long-extinct dinosaurs, explanations of the destructive power of volcanoes, and illustrations showing the different organs in the human body, I absorbed it all.

But astronomy held a special place for me, even then. Astronomy was different. Books about other parts of science told me all about how things worked, and listed all the things scientists had been able to figure out. It seemed that all the big questions had mostly been answered, and we now were merely sorting out the details. In contrast, books on astronomy seemed to focus not on what we knew, but on what we *didn't* know.

And what we didn't know was a lot. Astronomy grabbed my attention because there were far more mysteries than answers. What is "dark matter"? What's inside a black hole? Is there life on Mars? Science was supposed to be all about discovering things, and it seemed to my young mind that most of the discoveries still to be made were in astronomy.

So I decided very early on, maybe when I was only about

five years old, that I wanted to be an astronomer. It started with a fantastic book that my parents gave me, called *Album of Astronomy*. This book (which I still have) took the reader on a stunning tour of the solar system, the Milky Way Galaxy, and beyond. From the intense heat of Mercury and Venus and the fiery nuclear fusion reactions that power the Sun, to the glittering vast majesty of distant spiral galaxies and the unimaginable beginning of everything in the Big Bang, there were things out there incredibly foreign and alien to all my entire life experiences to that point. I was hooked.

As I got older, my fascination with the heavens only grew. In third grade I wrote a textbook on astronomy for my school library. When I was 12, I used my pocket money to buy a telescope, and then stayed up all night to watch Halley's Comet. And I begged my high school teachers to spend less time talking about chemistry or geology, and more time on astronomy.

Decades later, it's a huge thrill that astronomy is what I actually now get to do for a career. It's a thrill because what first attracted me to this topic as a child turned out to be true. Astronomy is still the unexplored frontier, and mind-blowing discoveries are still made all the time.

But there are two things about being an astronomer that did not turn out the way I expected—two things I didn't truly appreciate as a child, that only became apparent once I began my career. It's because of these two realizations that I've written this book.

Finding stars or understanding them?

When I was young I very much hoped and expected that my main activity as a professional astronomer would be to discover new stars.

And I have indeed been able to discover a few new stars during my career. I remember vividly the moment when I first discovered a star, in 1994. I was sitting at my computer screen looking through my data, and there it was—an object that no other human in history had ever gazed upon. I sat quietly for a few minutes, alone with my discovery, cherishing the fact that there was something out there that only I knew about.

But inevitably astronomy and science are all about sharing your ideas and discoveries. I emailed my colleagues to tell them what I had found, and in time we published our results in a scientific journal, for others to consider and study further.

The star I discovered is now known as "PSR J1024-0719," and is approximately 1,700 light-years (10,000 trillion miles) from Earth in the constellation Sextans. It's far too faint to see with the naked eye, but sits high in the sky in the evenings of March and April each year. While I have not ever studied or looked at PSR J1024-0719 again since that moment of discovery, it will always hold a special place for me as a mark of the beginning of my cosmic adventures.

However, despite the thrill of such discoveries, astronomy is not just about finding new stars.

It turns out to be quite easy to discover a star. Take any large telescope, point it in some random direction, and make a picture of the corresponding small patch of night sky on which the telescope is focused. The resulting image will be filled with stars, and almost all of them will be stars that no one has ever seen before. These stars will not have names, they will never have appeared in any catalogs, and almost nothing will be known about them.

As an astronomer, it's thus tempting to devote one's career to trying to find and catalog as many of these stars as possible. However, the reality is that many of these new stars will be rather plain and uninteresting, no different from the millions of stars that have already been named, cataloged, and classified. Sometimes a star turns out to be especially interesting or unusual (and I'll be introducing you to some of those stars throughout this book), but these special stars are extremely rare, and it isn't apparent that they are special or notable objects until astronomers have subjected them to considerable further study.

It's natural and understandable that I was excited when I made my discovery of PSR J1024-0719. But at the same time, it's important to appreciate that the goal of astronomy is not to catalog and collect, but to understand. Astronomers aren't motivated by the need to merely count stars or to determine their coordinates, but by the desire to determine what stars are and why they shine. Ultimately, astronomers seek to answer some of the biggest questions of all: Where did we come from? How will things end?

This is not to suggest that finding stars, measuring their properties, and putting them into catalogs are not hugely important parts of astronomy. Some of the largest and most important projects in astronomy have been massive efforts to carefully map and classify the objects in the night sky. These range from the landmark Henry Draper Catalogue of 225,300 stars completed in 1924, to the spectacularly successful Sloan Digital Sky Survey, which began in 2000 and has so far cataloged more than 500 million stars and galaxies.

However, the important thing to realize is that astronomers undertake these huge efforts because we can then use the completed catalogs to understand new things about stars, galaxies, and the cosmos.

So while, as a child, my ambition was to discover new stars, my goal now is to discover new things about how the Universe works. The latter approach far better captures what astronomers do, and why we're so dedicated to the cause. Like the young child who is always asking "Why?" we simply seek to understand the world around us.

Billions and billions

As a child I loved the idea that numbers go on forever. Every book has a last page. On every film, the credits eventually roll. And even if you read the biggest dictionary you can find, you will eventually run out of words to discover. But numbers don't have these limitations. A million, a billion, a trillion, a quadrillion . . . even after we run out of names for big numbers, they keep getting bigger.

But while I loved the idea of numbers beyond counting, I was intimidated by the fact that I couldn't really conceive what these numbers meant. What could "a billion stars" or "a million galaxies" actually mean? I wanted to be an astronomer because that way I could hold numbers such as these in my head and truly understand them.

But I now know that astronomers don't have some special way of looking at the sky. We cannot conceive a true understanding of just how big and complicated the Universe is any more than can anyone else. For example, while I was writing this book, American astronomers announced that they had recalculated the total number of stars in the observable Universe. Previous estimates had put this figure at 100,000,000,000,000,000,000,000 (100 sextillion) stars, but the new calculation showed that this number was more like 300,000,000,000,000,000,000,000 (300 sextillion) stars. Clearly the count of the total number of stars has tripled thanks to this new discovery, but both 100 sextillion and 300 sextillion are equally incomprehensible, as much to a professional astronomer as to anyone else. Our minds evolved to hunt for food, to avoid dangerous animals, and to interact with other human beings. Accordingly, we can think in terms of hours, months, and years, and can visualize distances in yards or miles, but numbers much bigger than this lose their meaning. The scale and scope of the Universe go far beyond what our minds are capable of processing.

Nevertheless, all is not lost. Mathematics and physics are incredible tools, because they allow us to study and understand the Universe, even when the numbers associated with astronomy are so far beyond our experience that they virtually have no meaning. What I now appreciate is that we can simultaneously be overwhelmed by the sheer scale confronting us, and yet still appreciate the extraordinary power, wonder, and beauty of the cosmos.

Extreme cosmos

In this book I want to convey to you just how far beyond our everyday experience the Universe extends in every imaginable way. But at the same time, I hope *Extreme Cosmos* will help you appreciate how truly remarkable it is that we can nevertheless make these measurements and write them down. What's more, in most cases we think we understand what these objects are, how they formed, and why they have their incredible properties.

In the chapters ahead, I've chosen 10 different concepts that we all experience on a daily basis: temperature, light, time, size, speed, mass, sound, electricity/magnetism, gravity, and density. For each of these there are extremes to our experience: We've all felt blazing heat and bitter cold; we've seen a jet plane speed overhead and watched a snail creep through the garden. In each chapter of this book I will start with these everyday ideas, but will then build a bridge to objects throughout the Universe with properties far beyond what any of us are capable of genuinely understanding. Sometimes I'll only be able to take you to one end of the spectrum: For example, for extremes of speed in chapter 5, we'll look at some of the fastest things in the Universe, but it's hard to ask or answer the question as to which stars are moving the slowest, because the vast majority of stars are moving much slower than our telescopes are capable of measuring. In other chapters it will make sense to explore both extremes: For example, in chapter 1, I'll be able to show you some of the limits of both hot and cold throughout the cosmos.

In writing this book, I want to convey the two things I have learned about astronomy in my journey from excited stargazing child to career scientist. I want to share with you that astronomy is so much more than finding new stars. And I want you to be assured that even though the numbers that describe the cosmos are utterly incomprehensible to our mere human minds, the remarkable beauty, diversity, and elegance that underpin these measurements are still deserving of our wonder and appreciation.

And finally, before we begin this journey, a caveat. In each chapter that follows, I've focused on a particular type of measurement, and tried to be as definitive as possible as to how the Universe pushes such measurements to the extreme. But definitive is not a term that's easily applied when studying the cosmos. For example, I can describe to you the deepest known note in the Universe (for the answer, see chapter 7). But I can't tell you with certainty that this is assuredly the deepest note to be found anywhere in the Universe, because there's so much of the Universe that we are yet to look at. Instead, all I can offer is our findings based on the objects that astronomers so far have been able to study.

What's more, sometimes the measurements involved are rough and uncertain. Later I suggest what might be the heaviest black hole in the Universe (see chapter 6). However, the masses of black holes are necessarily very crude, and my claims are limited by this uncertainty in our data.

Finally, astronomy is a dynamic, burgeoning field. New discoveries are made on a daily basis, and records are inevitably shattered. Even in cases where I can write with confidence and certainty, such as for the case of the most luminous object ever seen in the Universe (see chapter 2), an even more extraordinary object will no doubt soon come along to surpass it.

I was first attracted to astronomy because there was so much still to understand. Decades later, I remain thrilled by the realization that there is an endless road of discovery and delight ahead.

1

EXTREMES
OF TEMPERATURE

As planets go, Earth is a pretty hospitable place. It is, after all, teeming with life-forms that need temperatures to be neither too cold nor too hot. But anyone who has visited the Australian desert in summer, or spent a winter's night in Canada, will know that even our "just right" home planet is capable of conjuring up an incredible range of surface temperatures, well beyond the narrow comfort zone that we fragile humans can comfortably handle. Earth's extremes range from 136°F recorded in 1922 in Al 'Aziziyah in Libya, down to the bone-chilling −128°F in 1983 at Vostok Station in Antarctica. And of course most of the Earth's interior is much hotter than anywhere on the surface, while some parts of the Earth's atmosphere are far colder.

But the harshest climates that the Earth can offer are nothing compared to what is found elsewhere in the Universe. Throughout the deep reaches of the cosmos, there are places trillions of times hotter than the hottest sauna, and other places so cold they make Toronto on New Year's Eve seem like a beach picnic.

Hot and hidden

Let's first think about what "hot" and "cold" really mean.

Matter in its usual forms (solid, liquid, or gas) is made up of atoms and molecules. In a solid, the atoms or molecules are all held rigidly in place like interlocking pieces of a jigsaw puzzle. In a liquid, particles can move around but generally still remain clumped into large groups. And in a gas, every atom or molecule acts independently, and is free to travel wherever it desires.

However, what solids, liquids, and gases have in common is that their constituent atoms or molecules are all endlessly quivering and shaking. In a solid, each particle never moves far from its designated location, but every atom nevertheless jitters back and forth. (Imagine a jigsaw puzzle where the pieces don't fit together perfectly; each piece has room to move back and forth, while still always remaining in position.) In a liquid or gas, particles dance crazily in random directions, like an out-of-control bumper car.

What temperature measures, on a microscopic scale, is the speed of these jiggles and vibrations. Regardless of whether something is a solid, liquid, or gas, it can potentially be found in a state where these random motions are slow and gentle, or in a state where they are maniacally fast. Slow movement of the atoms or molecules corresponds to the object being cold, and rapid movement means it is hot. Cool something down, and the individual particles pare back their motions to a gentle waltz. Heat the object up, and they launch into a frenzied tarantella.

What this then means is that there's really no upper limit in temperature—if you keep heating something up, the particles inside will keep bouncing around faster and faster. With this picture in mind, we can now ask the question: How hot can the Universe get?

Let's start with the Sun, a giant burning ball of gas, so hot and intense that it's not safe to even look at. The Sun has a surface temperature of around 9900°F, which is hot, but not unimaginably so. The Sun's surface is about five times hotter than a candle flame, or a little more than double the temperature of a flame in a blowtorch. This is hot enough to melt tungsten, but not hot enough to boil it.

But there are other stars much hotter than the Sun.

We all know that if you heat something up, it glows. A poker in a fire shines a dull orange or red, while a conventional (incandescent) lightbulb works by heating up a tungsten filament to several thousand degrees so that it glows yellow or white. These are special cases of a universal process first properly explained by German physicist Max Planck: Virtually every object (whether on Earth or in space) radiates light, and the color of this light is tied to the object's temperature.

We can see this effect, known as "Planck's law of black body radiation," in action whenever we look at the different colors of stars. Our Sun is a reasonably average star. Its surface temperature of 9900°F results in a yellowish light, just as Planck's equations predict.

Betelgeuse, a bright star in the constellation of Orion, is much cooler, about 6900°F, and so even to the naked eye has an easily identified red hue. The brightest star in the night sky, Sirius (also known as the "Dog Star"), has a surface temperature of about 18,000°F, which gives it its bluish tinge.

But there are other stars, invisible to the naked eye, which are far hotter than Sirius. As we'll see a little later in this chapter, the real action is happening deep within a star's core, where the fury of nuclear fusion generates all a star's heat and light for up to billions of years. But when a typical star finally exhausts all its fuel, it puffs off most of its outer layers into a slowly expanding shell of gas, exposing the central core. This core, a small dense ball of helium, carbon, and heavier elements, is no longer burning any gas via nuclear fusion, but is still incredibly hot. This dying ember, known as a "white dwarf," is now among the hottest stars in the Universe, so hot that it lights up the surrounding shroud of expelled gas to form an exquisite glowing cloud known as a "planetary nebula."

So just how hot is a newly formed white dwarf? The current record holder sits at the heart of a beautiful planetary nebula. This glowing gas cloud, referred to by astronomers as "NGC 6537" but more commonly known as the "Red Spider Nebula," is about 2,000 light-years away toward the constellation of Sagittarius. (One light-year is the distance you can travel in one year if you move at the speed of light, a total of just under 6 trillion miles. So 2,000 light-years is around 12,000 trillion miles!)

Throughout the 20th century, the central white dwarf in the Red Spider Nebula eluded detection, and its temperature remained unknown. There are two reasons why such stars are so hard to see. First, they are tiny objects buried at the very centers of glowing, luminous, surrounding clouds. Often the brightness and complexity of a planetary nebula hides its central star from view.

But the other reason is that, paradoxically, the star's extreme heat itself makes the star almost invisible. As we saw above, Planck's law of black body radiation dictates that an object's temperature determines its color. Sirius, with its surface at a temperature of 18,000°F, is so hot that it glows blue.

What happens if a star is even hotter than blue Sirius? In such cases Planck's law still applies, but the resulting glow will be of a color beyond the range to which our eyes or ordinary telescopes are sensitive. In particular, objects much hotter than Sirius will glow in ultraviolet or X-ray light. Different temperatures, and their connection to color through the law of black body radiation, reveal that seemingly distinct phenomena such as ultraviolet light and X-rays are really just parts of the broad electromagnetic spectrum. The electromagnetic spectrum describes a whole range of different colors, well beyond the sliver of light that we can see with our eyes.

So white dwarfs are buried deep within their planetary nebulas, and are so hot that they don't emit much visible light, but instead radiate mainly in the ultraviolet and X-ray parts of the spectrum. It's thus not too surprising that the superheated star at the center of the Red Spider Nebula remained unseen for many decades. That situation finally ended in 2005, when Minkako Matsuura and his colleagues used the powerful Hubble Space Telescope, located in orbit above the Earth's atmosphere, to identify a tiny speck of light corresponding to the white dwarf at the heart of the Red Spider. In this and subsequent studies, astronomers have been able to make a precision measurement of the star's color, and then have used Planck's law of black body radiation to calculate its temperature.

The results are astonishing—the surface temperature of the star at the center of the Red Spider Nebula is an incredible 540,000°F, more than 50 times hotter than the Sun, and 30 times hotter than mighty Sirius.

This amazing star, with its extreme temperature and the spectacular glowing nebula that surrounds it, is of more than mere academic interest. For in gazing at the Red Spider, we are seeing our future fate. Around 5 billion years from now, the Sun too will run out of fuel, and will similarly shed its outer layers. All that will remain of our star and its solar system will be a beautiful planetary nebula, illuminated by an intensely hot white dwarf at its center.

The nuclear furnace

Stars may have high temperatures at their surfaces, but the fiery hell in their interiors is unimaginably hotter. Our Sun, an average and unremarkable star, provides the heat and light that make life possible. The Sun might be modest compared to some other stars, but it is still an impressive beast.

The Sun weighs about 2,200,000,000,000,000,000,000,000,000 tons (that's about 330,000 times heavier than the planet Earth) and is about 860,000 miles across. For the gas in the core of the Sun, about 39% of the total mass is hydrogen, 60% is helium, and the remaining 1% is made up of small amounts of carbon, oxygen, silicon, iron, and other heavier elements. (In fact, almost every element known has been identified in the Sun at some level. Even elements like silver, gold, and uranium are found in the Sun in trace amounts.)

When the Sun began its life 4.6 billion years ago, the composition of its core is thought to have been very different from what we see today—more like 72% hydrogen, 27% helium, and 1% everything else. This massive change in the core's composition over the Sun's lifetime, from 72% hydrogen at the beginning to 39% hydrogen now, provides a vital clue to the extremes going on in the Sun's deep interior. It tells us that the heat and light of the Sun come from nuclear fusion, in which hydrogen is continuously converted into helium, releasing large amounts of energy in the process. This is the same phenomenon that provides the devastating power of a hydrogen bomb, but on a far larger scale.

As its name suggests, nuclear fusion involves the nuclei of two hydrogen atoms sticking together. But this is not something that can happen easily, because hydrogen nuclei have positive electrical charges, and two positive charges will furiously try to repel each other when brought close together. It is only when two hydrogen nuclei can be brought close enough to essentially touch that they will then bind to make helium.

The trick is to bring the two hydrogen nuclei together as quickly as possible. If they approach each other slowly, they will have time to exert their repulsive forces on each other, and will stay apart. But aim them at each other at high speed, and their mutual electrical repulsion cannot slow them down enough to prevent a collision.

In the Sun, just as in a hydrogen bomb, this process is achieved by heating the hydrogen to extraordinarily high temperatures. At this high temperature, individual hydrogen nuclei fly around randomly at enormous velocities, making fusion possible via high-speed collisions. Calculations show that the temperature needed in a star's core to make this reaction proceed is about 9,000,000°F. At this temperature, all solids and liquids vaporize into gas, all molecules are broken apart into individual atoms, and electrons are then torn off these atoms to leave the central nuclei exposed.

While this temperature of 9,000,000°F is the minimum needed for a star to shine via nuclear fusion, the Sun is even hotter, with a core temperature of about 27,000,000°F! And as mentioned above, the Sun is an unremarkable star. Heavier stars generate even hotter temperatures in their cores—up to 90,000,000°F.

The period of time during which a star converts hydrogen into helium in its core is known as the "main sequence," and occupies most of the star's life. The Sun is at about the halfway mark of its main sequence phase, with another 5 billion years or so to go. For much heavier stars, which shine brighter and burn their fuel much more quickly, the main sequence can run its course 1,000 times faster.

When a star has converted all the hydrogen in its core into helium, the fusion reactions at the center shut off, and the star begins to run into the health problems associated with old age. In principle, all the helium nuclei now present in the core could also now fuse into even heavier elements. But since every helium nucleus has double the positive charge of hydrogen, the electrical repulsion between two helium nuclei is much stronger than for two hydrogen nuclei. Even at these extraordinary temperatures, helium fusion simply refuses to take place.

With its heat source extinguished, the star's core now begins to collapse under its own gravity, becoming smaller, denser, and even hotter. The star then becomes a "red giant" (more about this in chapter 4). Eventually the core of the star reaches around 180,000,000°F, a temperature at which helium nuclei now travel fast enough to collide and fuse to form carbon. The star now gets its "second wind," and enters a period of relative stability, known as the "horizontal branch" phase of stellar evolution.

But inevitably the helium too is used up. For a star like the Sun, this is now almost the end. The core will compress and heat up further, but there isn't enough mass to trigger any more nuclear reactions. A series of complicated convulsions begins, eventually leading to a wind that blows off the star's outer layers, producing a beautiful glowing planetary nebula like the Red Spider Nebula. What was the core of the star is left behind as a white dwarf, the incredibly hot, slowly cooling, dense ember that's all that remains of the central engine.

But for stars heavier than the Sun, the game is not yet over. Once all the helium has fused into carbon, the core heats up further, until at an unimaginable 1,000,000,000°F, carbon nuclei begin to fuse to form oxygen, neon, magnesium, and sodium. For the very heaviest stars, with masses more than eight to ten times that of the Sun, the primrose path has yet further to run. When the temperature of the star's core rises above 2,700,000,000°F, oxygen nuclei fuse to form silicon, sulfur, and phosphorus. And then at around 5,400,000,000°F, silicon fuses to form iron.

At this point a star's life is almost at an end, because iron is the most stable element in the Universe, and resists undergoing further fusion. With all nuclear reactions at an end, the largely iron core now collapses further, squeezed until it reaches a temperature of around 9,000,000,000°F. Yes, that's 9 billion degrees. The core then catastrophically collapses to form a ball of almost pure neutrons, only about 15 miles across. The outer layers of the star fall onto this newly formed neutron star and rebound, causing a vast supernova explosion that rips the outer parts of the star apart and blasts them off into space at high speed.

The millions and billions of degrees achieved at the centers of stars are temperatures nothing short of stupendous. But these are the intense conditions required to produce the light throughout the Universe from uncounted trillions of stars. The radiation and heat from the Sun that's essential for life on Earth is underpinned by extremes of temperature far beyond our comprehension.

Temperatures off the scale

The centers of stars are hot, but as you might expect, they've got nothing on the temperatures involved in the very earliest moments of the Universe. It's now been about 13.7 billion years since the Universe itself began in a "Big Bang," and as we'll see in a moment, things have now cooled down so that the empty space that fills most of the Universe is extremely cold. But if we run the clock backward in time, the cosmos gets very toasty indeed. If we go back to the Universe's very earliest moments, how hot do things get?

We'll start with the situation just one second after the Universe began. We can't make any measurements or observations of this era, but we can work backward from what we see now to make what we think are reasonably accurate calculations of the conditions at that time. When the Universe was one second old, the temperature everywhere was about 18,000,000,000°F, twice as hot as the core of a massive star at the last moment of its life. At this temperature, atoms could not yet exist: Their building blocks—protons, neutrons, and electrons—flew freely around in all directions, occasionally colliding, but with too much energy to ever stick together.

Let's step back further, to one-millionth of a second after the Big Bang. The Universe's temperature was now 18,000,000,000,000°F! The Universe was filled with small particles called "quarks," which were beginning to stick together to form protons and neutrons. (Quarks, which were given their odd name in the 1960s in honor of a line of poetry from James Joyce's *Finnegans Wake*, are fundamental constituents of matter. There are different types of quarks, and combining groups of three quarks in different ways creates particles like protons and neutrons.)

Does our understanding of our origins allow us to push back even further, to times earlier than one-millionth of a second post–Big Bang? Incredibly, yes. Just ten-trillionths of a second (0.00000000001 seconds) after the Big Bang, the temperature was around 18,000,000,000,000,000°F. The Universe was now a soup of elementary particles such as quarks, leptons (a family of subatomic particles that includes electrons), and gluons (particles that "glue" quarks together). Of these particles, 50% were normal matter, which at lower temperatures act as the building blocks for our bodies and the world around us. But the other 50% was "anti-matter"—similar particles, but with the opposite electrical charge.

Anti-matter can only exist under special conditions on Earth, because when a matter particle and its anti-matter twin come into contact, they annihilate in a flash of energy. But in the extreme conditions of the very early Universe, matter and anti-matter were too hot to interact with each other, and existed in almost exactly equal proportions.

Probably the earliest time we can meaningfully discuss is just 0.0000000000000000000000000000000000000000001 seconds after the Big Bang, when the temperature would have been 180,000,000,000,000,000,000,000,000,000,000°F. At earlier stages, the very concept of time becomes difficult to define, and we move into an era that goes beyond our current physical understanding. Furthermore, it becomes difficult to even say what we mean by temperature.

Why? Well, because as I explained earlier, the temperature of an object corresponds to how rapidly individual particles move and vibrate. But at the Universe's earliest stages, it is unclear whether particles as we understand them could even have existed. The Universe's extremes of temperature are not only beyond what we can comprehend, but are perhaps beyond what can ever meaningfully be measured.

The big chill

We have seen that there are almost no limits to how hot the Universe can get. As I explained earlier, as the temperature gets hotter and hotter, this corresponds to the motions of individual atomic or subatomic particles becoming increasingly frenzied. But let's now think about the opposite case, that of extreme cold.

As you cool something down, the atoms or molecules become increasingly sluggish in their motions. This immediately implies that there must be a minimum possible temperature, a degree of coldness at which every particle comes to a halt, and remains held in place.

Scientists first realized that there was a likely limit to coldness more than 300 years ago. This limit, now known to be at a temperature of −459.67°F, is known as "absolute zero," and is the lowest temperature that any object can ever have. In fact, the laws of physics forbid any object from ever completely reaching an exact temperature of absolute zero, but there is no practical limit to how close you can get to it. And with that understanding in hand, we're ready to handle the coldest places in the cosmos.

Space is cold. How cold? To be precise, the current average temperature of the Universe has been measured at −454.76°F, just 4.91° degrees above absolute zero.

While experiments in laboratories here on Earth have reached temperatures within a billionth of a degree of absolute zero, to reach these unbelievably frigid depths requires extremely complicated (and expensive!) equipment. Given that the Universe has no such equipment at its disposal, there's little doubt that space is impressively cold. Certainly it's much colder than any ordinary thermometer is capable of measuring. An old mercury thermometer, for example, freezes solid at −36°F. Even an alcohol thermometer stops working at around −150°F. So how do we measure the much lower temperature of the depths of space? And more significantly, what is stopping the Universe from cooling down all the way to within a hair of absolute zero?

Let's first go back for a moment to Planck's law of black body radiation, which relates an object's temperature to its color.

In the case of hot white dwarfs, we have seen that there are objects so hot that they no longer radiate much in the visible part of the spectrum, but instead emit most of their energy in ultraviolet light and in X-rays. In the same way, sufficiently cool objects glow in light that is similarly invisible, but on the other side of the visible spectrum. For example, the human body, at a temperature of 98.6°F, emits mainly at infrared colors, invisible to the unaided eye but seen easily with night vision goggles. Objects even colder than this will radiate in microwaves or radio waves.

In the late 1940s, scientists realized that if the entire Universe has a particular temperature, then Planck's law requires that it must glow at the color corresponding to this temperature. And indeed, in 1965, in an innocuous-sounding scientific article entitled "A Measurement of Excess Antenna Temperature at 4080 Megacycles per Second," two physicists, Arno Penzias and Bob Wilson, reported the accidental discovery of this all-pervading cosmic glow. The radiation that they detected had a color in the microwave part of the electromagnetic spectrum, corresponding to a temperature of −454°F. Penzias and Wilson received the 1978 Nobel Prize in physics for their discovery of this "cosmic microwave background" (or CMB, as it is usually known).

This discovery is certainly interesting, but why was it deserving of the Nobel Prize? Because the CMB, the weak universal glow of frigid space, is considered incredibly strong evidence that our Universe suddenly came into existence in the

Big Bang, 13.7 billion years ago. While we do not yet know just what triggered this Big Bang, our current understanding is that both space and time simultaneously began at that moment, and that the Universe has been expanding in all directions ever since.

Following the first extreme moments after the Big Bang that I described above, the Universe spent the next few hundred thousand years as an incredibly dense soup of protons, neutrons, and electrons. An observer at that time would have been shrouded in heavy fog, because every bit of light emitted by any object would have quickly collided with a nearby electron. But as the Universe expanded and cooled, it reached a point where it went through an abrupt change: About 380,000 years after the Big Bang, electrons and protons combined to form hydrogen atoms, and the fog suddenly cleared. At that moment the temperature of the Universe was about 4900°F, and correspondingly the Universe was suffused in a reddish-yellow glow. Over the 13.7 billion years that have followed, the Universe has steadily continued to cool, to reach its current chilly level. In the future, the temperature will slowly drop further, gradually approaching, but never quite reaching, the lowest temperature possible, absolute zero, at −459.67°F.

Astronomers have tried valiantly to come up with other explanations for what might fill all of space with the weak, ghostly glimmer of the CMB, but there seems to be only one possibility: The unimaginably cold temperature of space, at −454.76°F, is the fading archive of the Universe's fiery beginnings billions of years ago. This barely measurable glow is a vital key that has helped us understand the origins of the cosmos.

Colder than the cosmos

Are there any parts of space that are even colder than the CMB? The answer should be no, because even if such a situation could somehow naturally occur, the region would be bathed in the light of the CMB, which would heat it up to the same temperature, −454.76°F, as everywhere else.

But remarkably, there is at least one part of the Universe that is even colder than everywhere else. The Boomerang Nebula is another planetary nebula like the Red Spider Nebula, the result of layers of gas being shed by a dying star. As I intimated earlier, billions of years from now, when our own Sun runs out of fuel and nears the end of its life, it will produce its own nebula that might not look too dissimilar to the Boomerang Nebula. Such nebulas are relatively common. But like snowflakes, every one is different, reflecting the particular age, mass, composition, and other properties of the central star.

But the Boomerang Nebula is especially unusual. In this case, the dying star that created the nebula had an extremely strong wind: For the last 1,500 years of its life, the material in this wind was blasted out from the surface of the star at almost 400,000 miles per hour. And the rate at which the star shed material through this wind was immense, about 70,000,000,000,000,000 tons every second!

In 1990, astronomer Raghvendra Sahai realized that in such extreme objects, the wind from the star might not only travel at high speeds, but could also expand rapidly as it flowed outward. This rapid expansion can cause a dramatic drop in temperature, in the same way that sudden evaporation and expansion of coolant in your home refrigerator produces the cold temperatures that keep your food fresh. (What's going on here is the exact opposite phenomenon to squeezing a gas to make it hotter. If you've ever noticed how hot your bicycle pump is after you've inflated a tire, you'll know what I mean.)

A few years later, Sahai and his colleague, Lars-Åke Nyman, decided to test this idea on the Boomerang Nebula. They used a telescope in Chile to measure the temperature of gas in the Boomerang, and showed that it was indeed being "refrigerated" to very low temperatures. However, just how low was surprising: Analysis of the data showed that the gas in the Boomerang Nebula is at a temperature of −458°F, even colder than the CMB! Although the central star powering the Boomerang Nebula is very hot, the combination of a high-speed wind and rapid expansion has in this case produced the coldest naturally occurring place we know of in the Universe, with a temperature even lower than the extreme chill of surrounding space.

2

EXTREMES
OF LIGHT

I suffer from a medical condition known as "autosomal dominant compelling helio-ophthalmic outburst."

Don't worry, it's not as serious as it sounds. What it means is that whenever my eyes experience bright sunlight, I sneeze! (The acronym for the condition, of course, is "ACHOO.") ACHOO syndrome is extremely common, affecting about 20–30% of the human population. And for me, what it means is that before I go outside, I always need to check whether the Sun is shining. If it's cloudy, all is okay. But if it's sunny, I need to be sure to wear my sunglasses, lest I be overwhelmed by a fit of sneezing.

Sunlight might not have this particular impact on everyone, but to most people it matters each day whether the Sun is shining or not. For many of us, a sunny day might trigger us to apply some sunscreen, or to hang out our laundry on the line to dry, or to wash our car. Even the mundane tasks of everyday life have their schedule set by whether we can see the Sun.

More broadly, most life on Earth relies on the energy of the

Sun to survive, and has a behavior regulated by the 24-hour cycle of light and dark. Sunshine provides the light needed for plants and trees to photosynthesize. The darkness of the night is for many animals a time to sleep, knowing that they will not be found and eaten by the next species up the food chain. For every living creature on the surface of the Earth, bright light and deep darkness are a normal part of our daily existence.

But except on the occasional tour of an underground cave, we rarely find ourselves in almost complete, impenetrable darkness. And unless you look directly at the Sun, the light of day is not unbearably bright. But elsewhere in the Universe, there are extremes of darkness and light that go far beyond what we can ever expect to experience.

Dim and dusty

I grew up in a big city. While I spent hours in my backyard as a child looking at stars and memorizing the constellations, all the surrounding streetlights and house lights meant that my night sky was never especially spectacular.

So I'll never forget the first time I went out to the country. We drove all day, and arrived at our motel as the Sun was setting. We went into our room, unpacked our bags, and then watched some TV. A few hours later when it was completely dark, I stepped outside, and suddenly had to duck my head, instinct telling me that there was something right above the door that I was about to bang into. I looked up to see what it was, and discovered that it was the night sky! Stars, stars, and more stars, everywhere I looked, with the glorious Milky Way stretching in a massive arc right across the heavens.

Every time I get back to the countryside, the thing I look forward to most is reliving this moment. It's such a shame that this amazing tableau laid out on the sky, a display that people have wondered at for thousands of years, now barely peeks through the light pollution that swamps our towns and cities.

But if you too have escaped the city lights and looked up on a dark night, you will know that the night sky is anything but dark. The glitter of bright stars, the patterns of the various familiar constellations, and the broad glowing band of the Milky Way can be dazzlingly bright. The Universe might largely be a cold and empty place, but given the evidence of our own night sky, it would be reasonable to conclude that wherever you might be in the cosmos, it will never be dark.

But look more carefully. If the night is clear and if you wait long enough for your eyes to adjust, you will see inky patches of the sky where stars seem to be missing. One of the most famous of these is "the Coalsack," a region on the edge of the Southern Cross, a little smaller than your outstretched hand, in which no stars can easily be seen. An even more spectacular example, but too small to see without a telescope, is "Barnard 68," a tiny patch that looks like someone has taken a pair of scissors to the night sky and cut all the stars out.

The Coalsack and Barnard 68 are not places devoid of stars, but are dense gas clouds that block out the light from behind. Known as "dark nebulas" or "molecular clouds," they are the darkest regions of our Universe—their interiors are almost completely devoid of light.

Let's imagine that one of these clouds drifted through our part of the Milky Way, enveloping the Earth, Sun, and the rest of the solar system. In the direction from which the cloud approached, there would be a growing inky dark patch, eventually blotting out all the starlight from half the sky. But looking in the other direction, out to free space, we wouldn't notice any difference at all at first. The stars in that direction would seem just as bright as always.

After about 2,000 years (by which point we would have penetrated around 20% of the way into the center of the cloud), the half of the sky toward the cloud would remain totally black, but now the other half of the sky too would have started to fade. Over the centuries, the light from the various stars and constellations would have dimmed by about a factor of six—only about 150 stars would still be bright enough to be visible to the naked eye.

Wait another 2,000 years, and the remaining half of the night sky would fade by a factor of 20, leaving only 10 stars that we could see unaided. And if 2,000 years passed once more (a total of 6,000 years since our encounter with the cloud began), there would be no stars left at all visible with the unaided eye.

Finally, after 10,000 years, our solar system would be completely enveloped, and would now sit near the center of this enormous cloud. How dark would the sky be then? Thankfully, the light from the Sun would be largely unaffected. The days would seem normal, and we would also still be able to see the reflected sunlight that illuminates the Moon and planets.

But the rest of the sky would be utterly, totally void. The light from the rest of the Universe would be dimmed by a factor of 1,000,000,000,000. The brightest star in the sky, instead of being something that stands out easily to the naked eye, would be right at the limit of what the mighty Hubble Space

Telescope can detect. And all the other stars in the sky would be completely invisible to any telescope.

Of course, if humanity had arisen in such an environment, it's unclear whether we would even have developed such telescopes to search for starlight. After all, our curiosity about the night sky has been driven by our desire to better understand the stars. Without any stars to ever gaze on or wonder at, it's doubtful that astronomy would have been an important part of our history.

What makes these clouds so dark? Much to my mother's horror, the Universe is filled with dust. Not quite the soot, soil, and scraps of hair that might make up the dust in your living room, but tiny particles of silicates, graphite and ice, each much less than a ten-thousandth of an inch across. When a ray of starlight encounters such a dust grain, the light is either absorbed by the grain or scattered along a new, random trajectory. Gather enough of these dust grains together, and very little light can get through. Dark nebulas are especially dusty, and so prevent virtually all of the light from the rest of the Universe from penetrating into their interiors.

It makes me a little uneasy to think that there are places in our Galaxy that are so completely cut off from the splendor of the rest of the Universe. It's been many years since I was afraid of the dark, but this concept of utter, impenetrable blackness, a place in the Universe in which the rest of the cosmos is forever hidden from view, fills me with a sense of loneliness and isolation. However, without these dark, dusty nebulas, life on Earth might never have emerged.

This is because dark nebulas are incredibly important melting pots, and are one of the few places in the Universe where complex molecules can form. Molecules, of course, are the combinations of two or more atoms. They can be simple constructions like H_2O: two hydrogen atoms bonded to an oxygen atom to make a molecule of water. Or they can be unbelievably complicated constructions, like the hundreds of millions of carefully arranged atoms that form a single molecule of human DNA.

But let's first consider the most common molecule in the Universe, two hydrogen atoms joined together to make "molecular" hydrogen, which we write as H_2. The Universe is full of individual hydrogen atoms, but they almost never combine to form H_2. There are a couple of simple reasons for this.

First, as we will see in chapter 10, most of the gas in interstellar space is extremely rarefied, so the chances of two separate hydrogen atoms ever colliding and merging is extremely low.

Second, if two hydrogen atoms ever do happen to crash into each other, they need time to form a connection and become a single molecule, in the same way that you need to give superglue time to bond when you glue a broken teacup back together. But randomly colliding hydrogen atoms will immediately rebound like billiard balls; they are never in contact anywhere long enough for a single H_2 molecule to form.

So how does H_2 ever come into existence? The answer is dust. Dust grains in dark nebulas are not tiny little hard spheres, but pitted, dented, distorted blobs, resembling microscopic gold nuggets. There are lots of cavities and nooks on a dust grain, perfect for capturing and holding a hydrogen atom. And should a lone hydrogen atom bump into a dust grain, that's often exactly what happens—the atom does not rebound, but sticks to the dust grain, held in place by a naturally occurring hole or hook. The atom now travels around with the dust grain, and sometimes, eventually, another single hydrogen atom is captured in the same way. If the two atoms, both stuck to the dust grain, are close enough together, they can join together to form a new H_2 molecule! The molecule eventually becomes unstuck, and the hydrogen atoms float freely once again, but now paired together.

Dark nebulas aren't just vital for making molecules, but for preventing them from being broken back up again into their constituent atoms. All molecules are comparatively fragile, and can't survive for long in the harsh conditions of space. Even the weak light of distant stars is usually energetic enough to break apart most molecules into individual atoms. So not only are dark nebulas one of the only environments in the Universe in which molecules can form at all, but they are a place where these molecules can gently and gradually combine into increasingly more complicated forms, without fear of being shattered back into their components by starlight from outside.

Astronomers have thus been able to demonstrate that dark nebulas are stocked with all sorts of molecules. The most plentiful are molecules of hydrogen (H_2) and carbon monoxide (CO). However, a wide range of more complicated molecules have also been discovered, including nitrous oxide (laughing gas), acetone (the main ingredient in nail polish remover), ethanol (drinking alcohol), and also, perhaps inevitably, acetaldehyde (the chemical that produces hangovers).

We do not yet know how life began on Earth, but we suspect that it somehow emerged from a complex soup of organic molecules that appeared during the Earth's early stages. It now seems likely that many of these molecules were rained down onto the young Earth by a bombardment of asteroids, which carried as their payload the molecules formed in dark nebulas. It is thus the starless, almost totally dark environments of these dense clouds that are likely to be responsible for our very existence.

A vision splendid

Let's return to the beautiful starry sky as seen from Earth on a clear, dark night. This panorama, as glorious as it appears, is comprised only of the few thousand stars bright enough to be seen with the naked eye, and is just a tiny fraction of the hundreds of billions of stars in the entire Milky Way—let alone all the stars in the many billions of other galaxies out there.

However, some of the stars in the sky are not what they seem. Take the star known as "Omega Centauri," which is located 15,000 light-years away toward the constellation Centaurus. Omega Centauri is reasonably easy to see with the naked eye, and has been in star catalogs for thousands of years. But a careful observer will note that Omega Centauri is not a pinprick of light, but rather a small, fuzzy gray ball. Through a telescope the picture becomes clear: Omega Centauri is not a star, but a "globular cluster": a tightly packed group of more than a million stars.

We now know of more than 150 globular clusters in our Milky Way, although only Omega Centauri and two or three others are bright enough to see with the unaided eye. Globular clusters are remarkable for many reasons, but their immediate and most defining characteristic is that they are places where stars are found incredibly close together. A typical globular cluster might contain 500,000–1,000,000 stars, all within a region of space only 30 light-years across. This is an extraordinarily crowded situation—for comparison, within 30 light-years of the Sun, there are probably only about 500 stars.

If we were to look to the evening heavens from a planet inside a globular cluster, we would be confronted by an unbelievably rich and complicated night sky. For example, as viewed from Earth, the constellation of the Southern Cross (the smallest of the 88 officially recognized constellations in the sky) contains four bright stars (as represented on the New Zealand flag), plus a fainter fifth one (as included on the Australian flag). But if we were inside a globular cluster, a similarly sized patch of the sky would contain more than 1,000 stars! And this would be the same all across the sky, everywhere we looked.

The combined light of all these stars would be about the same as the full Moon, every night of the year. Clearly it would be difficult to come up with any stories or legends or mythology about such a sky, because it would be impossible to recognize any patterns or constellations—there would simply be nothing but stars and more stars, everywhere we looked. Even the glorious view of the overall Milky Way, the glowing white band that stretches across the entire sky on a dark night on Earth, might be hard to see, hidden by the dazzling glare of so many stars.

Globular clusters are the "mad scientists" of the Milky Way, because all sorts of otherwise impossible experiments become possible in their dense stellar environments. Ordinarily, in normal neighborhoods like that of the Sun, some stars are born in what are known as "binaries" (in which two stars bound by gravity orbit each other), but otherwise stars never get especially close together. Stars are relatively small, and the distance between them is enormous, so most stars go about their own business, unbothered by what might be going on next door.

However, in a globular cluster, the nearest stars are not your next-door neighbors, but fill every room of your house! Stellar interactions that simply would have no chance of ever happening in normal space become routine occurrences, and the results can be bizarre and complicated. As a star drifts around randomly inside a globular cluster, it can be captured by the gravity of another star passing by, and become part of a newly made binary. But just as often, a third star will wander through a binary system and will steal one of the pair of stars as its own companion, leaving the other star as a sudden singleton. Occasionally the third star will enter into a complicated long-term gravitational dance with the binary that it encounters, creating a new threesome that can persist for thousands of years until, perhaps, yet another star comes too close and breaks them apart again with its gravity.

Sometimes when a binary is broken apart by an interloper, one of the stars is ejected from the globular cluster at high speed, and flies off to join the rest of the Milky Way, never to return. At other times, two stars will actually collide and merge, forming strange, hybrid, Frankenstein stars such as "blue stragglers" and "Thorne-Żytkow objects." Astronomers have identified many examples of stars in globular clusters whose existence should be forbidden by the laws of stellar evolution—the explanation is almost always that in the crowded environment of globular clusters, almost anything can happen.

As well as being laboratories for unusual stellar experiments, globular clusters are also key to our overall understanding of stars. Astronomers often compare two stars and try to explain their different properties (for example, one star might be hotter or a different color from another). The main effect underlying these different properties is thought to be how heavy each star is, since massive stars are much hotter and more luminous than light ones. However, if you simply picked two stars at random, and concluded that the bright one was heavier than the faint one, you would almost always be wrong, because there are many other complicating factors that also need to be taken into account.

Foremost, we need to know how far away each star is, since a feeble star can appear bright if it is very nearby, while a powerful star can appear faint if it is far away. We also need to find out how old each star is, since stars gradually brighten as they age. And we need to determine each star's precise chemical composition, since the presence of small amounts of elements like carbon and oxygen can substantially affect a star's energy output. In many cases, it can be incredibly time-consuming to make these measurements, especially if a large number of stars are involved. In other cases, especially when the stars being studied are faint, the required measurements simply aren't possible using current technology.

A wonderful solution to these problems, and thus to getting a better overall insight of how stars work, is to compare different stars inside the same globular cluster. This is because, according to our current best understanding, the million or more stars in a globular cluster were formed at the same time, out of the same original cloud of interstellar gas. Therefore, every star in a globular cluster is the same age, has a virtually identical chemical composition, and because they are in such a small, tightly packed ball, is essentially at the same distance from Earth. If we see two stars in the same globular cluster that differ in their brightness, color, or temperature, then (excluding the occasional oddballs produced by mergers and collisions as mentioned above) the only possible explanation is that one star is heavier than the other. Globular clusters are thus essential tools for calibrating our understanding of how the properties of stars depend on their masses, providing a crucial component in our knowledge of how stars work.

Globular clusters were also the gateway to our modern view of the Universe. A hundred years ago, we knew that we lived in a galaxy called the Milky Way, and that the Milky Way was a large, flattened disc, with a diameter much larger than its thickness. All of this is consistent with our current understanding. However, at that time it was thought that the Sun and our solar system sat somewhere near the center of this large disc, with the edge of the disc roughly 20,000 light-years away in all directions.

This picture was shattered in 1918, through studies of globular clusters by the American astronomer Harlow Shapley. There were about 70 globular clusters known at the time, and Shapley developed a way of estimating a rough distance to each of them. Combining this with their positions in the sky, he was able to develop a three-dimensional model of the locations of globular clusters throughout the Galaxy.

Since globular clusters are relatively large, prominent agglomerations of stars, it was expected that they would be evenly spread throughout the Milky Way. If the Sun was indeed near the center of the Galaxy, then a three-dimensional map of globular clusters should show them scattered approximately equally in all directions. But Shapley's surprising discovery was that the locations of globular clusters are extremely lopsided—they are clustered in a roughly spherical distribution, but the Sun is not anywhere near the center of this sphere! This provided dramatic evidence that the Sun sits at an uninteresting and unimportant location, far from the Milky Way's center. (The latest measurements indicate that the disc of the Milky Way is about 100,000 light-years across, with the Sun roughly halfway from the center to the edge.)

Until the end of the Middle Ages, people thought that the Earth was the center of the Universe. Our modern picture is the opposite extreme: We live on a small planet, orbiting an ordinary star, hidden in the quiet suburbia of the Milky Way, which is itself a typical galaxy located in an unremarkable part of the Universe. Although there have been many discoveries that have gradually led us from the simple Earth-centered model of the cosmos to the complicated cosmology we study today, the story has been dominated by two seismic shifts in our understanding. The first was in 1543, when Nicolaus Copernicus published his incredible treatise "De Revolutionibus Orbium Coelestium" ("On the Revolutions of the Celestial Spheres"), which dramatically demoted the Earth, and moved the Sun to the center of all creation. But the second great moment was in 1918, when Shapley published his paper pragmatically titled "Studies Based on the Colors and Magnitudes in Stellar Clusters. VII. The Distances, Distribution in Space, and Dimensions of 69 Globular Clusters," demonstrating that even the Sun was not especially important or central.

So although the view of the night sky from Earth might be

a little bare and disappointing compared to what might be offered elsewhere, keep an eye out for those fuzzy gray balls that correspond to the brightest globular clusters. These are the telltale clues to our place in our vast Universe.

Ending in a blaze of glory

Sitting in your backyard—or in fact anywhere on the surface of the Earth—the Sun is unbearably bright, brighter than anything else in the sky. But this is only because it is so close—about 270,000 times closer than the next nearest star. If we viewed the Sun from 50 light-years away (a relatively small distance, still in our local neighborhood), it would be a faint star, barely visible to the naked eye. As viewed from 50,000 light-years away (a large distance, but still well within our Milky Way), the Sun would be visible only with a large telescope. And from 2 million light-years away (the distance to the nearest large galaxy, but only a tiny, tiny fraction of the way to the most distant galaxies we can see), the Sun would be far too faint to ever be detected.

There are many stars much more luminous than the Sun, and even if they are millions of light-years away, some such stars are still bright enough to be studied individually through powerful telescopes. But what is the most luminous star in the Universe? And how far away can it be seen?

As we saw in chapter 1, stars weighing more than eight to ten times the mass of the Sun end their lives suddenly and catastrophically, in enormous explosions called "supernovas." For a few days, the light from a supernova explosion can be a billion times more powerful than the Sun—so much so that a supernova can easily outshine the combined light from all the billions of stars in the galaxy that hosts it.

When a supernova occurs in our own Milky Way, it can be seen easily with the naked eye for months after the explosion, visible even in daylight if it is relatively nearby. The last time such an event appeared in the skies was Kepler's Supernova, which occurred in October 1604 in the constellation of Ophiuchus.

It's also easy to see supernovas in nearby galaxies. The most famous example occurred in February 1987, when the light reached Earth from a supernova explosion in the Milky Way's nearest neighbor, a galaxy known as the "Large Magellanic Cloud," about 170,000 light-years away. This was much more distant and much less spectacular than Kepler's Supernova of 1604, but was still clearly visible to the naked eye. Indeed, a pivotal moment in my own path toward becoming a professional astronomer was as a teenager standing in my backyard in the suburbs in February 1987, staring up at the constellation of Dorado, and seeing a new star shining that had not been there a few nights earlier. I was able to spot it every night, until it gradually faded away over the next few weeks. It was a huge thrill to know that the light that was entering my eyes had begun its journey in an unimaginably violent detonation 170,000 years earlier. For me, this captured the excitement and majesty of astronomy, and sharpened my focus toward pursuing a career studying the heavens. (And indeed for my PhD thesis 10 years later, I was lucky enough to study this very same supernova, as the stellar debris continued to hurtle outward into space.)

Using even a relatively small telescope, it is easy to see a supernova occurring in a galaxy millions of light-years away. It would be a frustrating experience though to simply pick one galaxy and look at it every night waiting for an explosion to occur, because in any one galaxy a supernova is a relatively rare event. In a typical galaxy, it is thought that on average a supernova occurs only about every 50 years, so staring at one galaxy for this length of time would not be a very practical experiment.

(Surprisingly, we haven't seen a supernova in our own Milky Way since Kepler's Supernova of 1604. Our Galaxy *should* be producing a supernova every 50 years just like other galaxies, so either we've been spectacularly unlucky, or many of the more recent supernova explosions have been hidden behind the obscuring dark nebulas that I mentioned earlier, which are scattered throughout the Milky Way.)

Fortunately, we don't have to wait anywhere near this long, because there are a vast number of galaxies we can study. If we could continuously watch 50 galaxies, each with a supernova occurring once every 50 years, then we would only need to look for a year before we could expect to discover a supernova. Expand this project to 2,000 galaxies, and we could expect to see a supernova explode almost every week. And indeed this is what supernova hunters do—every night, they point their telescopes at dozens of galaxies, comparing the resulting pictures to those taken of the same galaxies a few weeks before, looking for new pinpoints of light that would reveal the death of a star. Over the course of a year, many thousands of galaxies can be searched in this way, with spectacular results. In the year 2011, such efforts led to the discovery of 290 supernovas all over the sky. In our modern picture, supernovas are not the naked-eye events recorded once every few centuries in our history books, but events that go off like popcorn, all over the Universe.

Supernovas are a hot topic in astronomy (as evidenced by the intensive efforts that have discovered so many of them). One reason for this is that supernova explosions have played a major role in creating all the heavy elements in the Universe, including many of the atoms in our own bodies. As we saw in chapter 1, the giant fusion reactors at the cores of stars convert hydrogen into helium, then helium into carbon, then, in the most massive stars, carbon into even heavier elements. This sequence ends when silicon fuses to form iron, because even at the extreme temperature and pressure of a star's core, iron will not fuse any further. And yet all around us and inside us are elements such as gold, tin, iodine, and uranium, all of which are much heavier than iron. What created these elements? One of the few answers available is the extreme conditions of a supernova, which forge heavy elements that can't be produced by any other means. The explosion not only creates these new elements, but scatters them out into space. When an interstellar gas cloud later coalesces into a new star and its planets, it does so having been seeded with the heavy elements formed in a supernova. We are not just detached observers of the Universe, but are very much part of it, for we are all made of the ashes of ancient supernovas.

Supernova explosions, at least temporarily, are some of the most luminous events in the Universe. But we now know that there is a special, rare class of exploding star that far outshines mere "ordinary" supernovas. These events are known as "gamma-ray bursts," because they are detected as a sudden flash of gamma rays—an exotic form of light with the smallest wavelength and most energy of the entire electromagnetic spectrum. Gamma-ray bursts are seen about once per day, and are so powerful that they can be seen relatively easily from anywhere in the Universe.

So just how intense are gamma-ray bursts? A typical gamma-ray burst is 100–1,000 times more luminous than a supernova explosion! At one point this was attributed to some stars occasionally exploding with vastly more energy than normal. But most astronomers now think that this is an illusion— gamma-ray bursts are in many respects ordinary supernova explosions, but differ in that they additionally generate two intense, narrow jets of radiation directed along the north and south poles of the exploding star (the precise reason for the production of these jets is still not completely clear, but current evidence suggests that this happens only to the very heaviest stars). Viewed from almost all possible angles, these jets are invisible, and the gamma-ray burst appears more or less as a normal supernova. But perhaps one in a thousand gamma-ray bursts is oriented so that one of its jets is pointed right at us. It is in these cases that we see the characteristic intense bright flash of gamma rays. In some instances, if we look carefully after the gamma-ray burst has finished, we can see the fading supernova underneath, confirming that these incredibly powerful events are special alignments of exploding stars.

If you happen to be looking at the right position in the sky at the moment a gamma-ray burst occurs, the gamma rays themselves will not reach you because they will be blocked by the Earth's atmosphere (and can only be detected by telescopes launched into space), but what you will see is a flare of normal, visible light. In the minute or so that a gamma-ray burst is flaring in this way, it is producing enough light to comfortably outshine every other object in the entire Universe. The current record holder for the most luminous object in the Universe? A gamma-ray burst in the constellation of Boötes known as "GRB 080319B," which was easily visible to the naked eye for about 30 seconds on March 19, 2008.

Normally, the most distant astronomical object able to be seen without binoculars or a telescope is a galaxy known as "Messier 81," which is about 12 million light-years away in the constellation of Ursa Major, and just faintly visible to a stargazer with a very sharp eye as a small gray patch. However, GRB 080319B shattered this mark, for it was a naked-eye object whose light had traveled 7.5 billion light-years. While many aspects of gamma-ray bursts are still unclear, one thing that cannot be questioned is that gamma-ray bursts are by far the most powerful lights in the cosmos.

3

EXTREMES
OF TIME

We humans have great difficulty understanding the passage of time. To young children, any span of time longer than a few minutes seems like an eternity (hence the inevitable question: "Are we there yet?"). As we get older, we quickly become comfortable with the relative durations of hours, days, weeks, and years, but even then, our perception of time ebbs and flows at different points in our lives: Some years crawl past, while others seem to race by.

Quantifying the milestones of a typical human life span is hard, but dealing with cosmic time scales is much harder. In trying to understand the vast eons over which the Universe has existed, we're again reduced to the mind-set of a young child, who knows how long one day is, and who can imagine two days in succession or even three. But from a child's perspective, trying to imagine 365 of these days strung in a row is an impossible, frustrating exercise.

And so it is with the cosmos. Most things in the Universe happen incredibly slowly by human standards, as evidenced by the fact that the constellations and stars described by the ancient Greeks and Egyptians thousands of years ago are essentially unchanged today. The Earth and Sun are 4.6 billion years old, themselves relative youngsters in a Universe that began about 13.7 billion years ago in the Big Bang. (Our understanding is that the concept of time itself began at the moment of the Big Bang, so if there's an answer to the question of what happened beforehand, it's unclear if it would be an answer we could comprehend.)

Scientists can calculate these numbers and write them down, but we can't really imagine what they entail. So with the caveat that we can't hope to properly understand the magnitude of the numbers involved, astronomy can nevertheless reveal to us the unbelievable extremes of time that the Universe can produce.

The iron clock

Sydney is the oldest city in Australia, having been founded with the arrival of British soldiers and convicts in 1788. I was born and bred in Sydney, and often wonder what my hometown might have looked like in those earliest days. However, I am left to use my imagination, because that early Sydney is almost completely gone, swept away by the tides of progress and construction. If you wander through the center of Sydney, near the iconic harbor, the oldest building you can find—a modest sandstone cottage—dates only back to 1816. If you broaden your search to the suburbs, many miles from the city center you will eventually find two older buildings dating back to the 1790s. Few other traces of the original Sydney remain.

In many respects, attempting to trace the history of the Milky Way is similar to trying to picture the early days of Sydney. We know that the Universe is 13.7 billion years old (give or take a few hundred million years or so). According to current calculations, the Milky Way is not much younger, having begun to take shape more than 13 billion years ago.

What our Galaxy looked like back then, and how it eventually took on its current appearance, is a major area of current astronomical research. There is even an entire sub-field of astronomy called "galactic archaeology," aimed at using the Galaxy's "fossil record" to understand its history. One thing we know is that few clues are going to be found close to home: Our best estimate, based on radioactive dating of meteorites, is that our own solar system is a mere 4.6 billion years old. Unimaginable eons of galactic evolution had taken place long before the Earth and Sun were even a twinkle in the cosmos's eye.

The challenge, just as for historians studying Sydney, is that so much from those early times has been changed, destroyed, or rebuilt. The Milky Way is a dynamic, violent, energetic place. As we have seen in previous chapters, stars end their lives, often in spectacular fashion. The interstellar clouds that this produces then re-form into new stars, and the process begins again. Meanwhile, the spiral-arm pattern that defines the Milky Way sweeps through gas and dust, churning and stirring and heating it. Our Milky Way is a reasonably large galaxy, so while all this other activity is going on, it also ensnares other, smaller galaxies with its gravity, engulfing them, digesting them, and mixing in their stars with its own.

All this turbulent activity means that no obvious record remains of the Galaxy's early history—we are unlikely to ever find the astronomical equivalent of the Roman Forum or the ruins of Machu Picchu, which we could then use to reconstruct the past. Our only hope is that of the hundreds of billions of stars in the Milky Way, perhaps there are a few survivors of our Galaxy's early days. Like a fragment of china or a rusted spoon in an archaeological dig, these ancient artifacts, our Galaxy's oldest stars, could provide crucial clues to previous eras.

To understand how we might search for such stars, we need to return to the process of nuclear fusion through which stars produce their heat and light, as was discussed in chapter 1. As we saw, stars derive their energy by fusing hydrogen into helium, then helium into carbon, and so on all the way up to the formation of iron. For any given star, as I'll explain below, this step-by-step process of atomic fusion works like a kind of natural clock.

As we saw in chapter 1, the Sun began its life made up of about 72% hydrogen, 27% helium, and 1% all the other elements like oxygen, carbon, and iron. Since hydrogen and helium are by far the dominant elements in the Universe, and since all the other heavier elements are found only in comparatively tiny proportions, astronomers rather whimsically define three types of normal matter in the Universe: hydrogen, helium, and "metals." In this context, the word "metal" does not refer to metallic everyday materials like silver and gold, but to everything else in the Universe besides hydrogen and helium! So a simple way of describing the chemical composition of the Sun is that its "metallicity" at birth was 1%.

The metallicity of a star is the cosmic clock that allows us to search for the oldest stars. The key point underlying this claim is our detailed understanding of the Big Bang and its aftermath, which tells us that before the first stars formed, the composition of the Universe was 75% hydrogen, 25% helium, 0.00000001% lithium and beryllium, and with no heavier elements at all. If the very first stars were formed out of this material, then their metallicity was virtually zero. For most of their lives, these stars would have fused hydrogen to form helium, but no metals would have been formed, and their metallicity would have remained at zero. Toward the end of their lives, these first stars would have fused helium to make carbon, and perhaps would have gone on to make some heavier elements also. At the ends of their lives, they would have had a small amount of metallicity.

When these stars then died, and the next generation of stars began their lives, they would have formed out of new clouds of gas "polluted" by these metals, and so would have started with a small metallicity. Eventually they too would make their own carbon, oxygen, and other heavier elements, and so would die with a small but higher metallicity.

And thus we have a cosmic clock. Newborn stars in our neighborhood, formed out of the detritus of many generations of star birth and star death, have a high metallicity. Middle-aged stars like the Sun, full of impurities but not as many as stars born today, have an intermediate metallicity. And ancient stars, if any survive from the Milky Way's earliest days, should have almost no metallicity at all.

For some time, the search has been on for such metal-poor stars, ancient survivors from a bygone area when the Galaxy was young. Astronomers conduct such a search through a detailed forensic analysis of starlight. All stars emit a range of colors, but careful examination shows that particular precise shades are dim or missing. This is because metals in the star's atmosphere block out particular colors. The exact pattern of missing colors is, like a fingerprint or DNA, unique to a particular star, and allows us to precisely calculate what fraction of metals any given star contains (again, remembering that a metal in this context means any element other than hydrogen or helium).

Using this technique, known as "spectroscopy," astronomers have had considerable success finding stars that are unusually metal-poor, and hence extremely old. An especially useful guide is the level of the element iron inside a star. As a point of reference, about 0.1% of the Sun is iron. This sounds small, but is a fairly typical iron abundance (and indeed by everyday standards this is still a lot of iron—equivalent to about 400 times the mass of the Earth!).

The most metal-poor star currently known is "SDSS J102915+172927," which is about 4,000 light-years from Earth in the constellation Leo. SDSS J1029 had previously been listed in star catalogs, but had had little attention paid to it. However, this all changed in 2011, when German-based astronomer Elisabetta Caffau and her team found that this star had far fewer metals than any other star ever discovered. In particular, according to Caffau's measurements, only about 0.00000003% of SDSS J1029 is iron, which is about 100,000 times smaller than the amounts seen for the Sun! Even our little planet Earth contains 100 times more iron than found in SDSS J1029.

The tiny amount of iron in SDSS J1029 tells us that, when this star was born, the metallicity clock had barely started ticking. While it is difficult to calculate an exact age for SDSS J1029 (and it is impolite to ask), a reasonable estimate is that this star is about 13 billion years old—vastly older than the Sun and almost every other star in the Milky Way. Like a wrinkled super-centenarian at a nursing home, SDSS J1029 is a final survivor of a bygone age. Astronomers are now trying to listen carefully to its faint, quavering voice, eager for stories of the way things used to be long ago.

SDSS J1029 is the oldest star known, but the hunt is on for stars that were born even earlier. For while the amount of iron in SDSS J1029 is comparatively tiny relative to much younger stars like the Sun, this still corresponds to 20 million trillion tons of iron. All of this iron must have been formed in other stars, before SDSS J1029 was born. And the even earlier generation of stars that made this iron in their stellar furnaces perhaps began their lives with no iron at all, having formed out of the primordial gas clouds of hydrogen and helium left over from the Universe's beginning.

One of the major efforts of modern astronomy is to find these first stars, known as "Population III." (SDSS J1029 is part of "Population II," while a relatively metal-rich star like the Sun is a member of "Population I." Each population reflects many successive generations of stars, so that the Sun's distant ancestors were Population II stars, and these in turn were preceded even longer ago by Population III stars.)

It is probably fruitless to look for Population III stars in the Milky Way, because such stars are thought not to have lived for very long, just a few million years. If there were ever any Population III stars that laid the foundations for what is now the Milky Way, they are long gone.

Our only way of finding Population III stars is to look back in time, by studying the light from incredibly distant stars in other galaxies. But before we proceed, this probably needs a little more explanation.

Light does not travel instantaneously, but moves at a finite speed of 186,282.397 miles per second (which is around 670 million miles per hour, or about 6 trillion miles per year). Because it takes time for light to travel from a heavenly body to our eyes or to our telescopes, we can never see how a planet, a star, or a galaxy looks right now, but instead we view the object as it appeared when the light began its journey to us.

This effect is ever present, but for many objects in the night sky, it hardly matters. For example, it takes light almost 8.5 minutes to travel the 93 million miles from the Sun to the Earth. So when we study the Sun, we don't see how it appears right at that moment, but how it looked 8.5 minutes earlier. Sirius, the brightest star in the night sky, is about 8 light-years away, so if we look up and see Sirius at night in the year 2012, we are actually seeing how it appeared in the year 2004. But since the Sun and most stars change very slowly, on time scales of years, centuries, and millennia, we normally don't bother taking into account the fact that what we see today is a little out of date.

However, look far enough away, and this starts to make a difference. In chapter 2, we talked about the most luminous object in the Universe, a gamma-ray burst that, for 30 seconds on March 19, 2008, was bright enough to be seen with the naked eye. The distance to this gamma-ray burst was 7.5 billion light-years, meaning that the explosion did not occur on March 19, 2008, but happened billions of years earlier, before the Earth or Sun even existed! Right throughout the history of our solar system, the light from this incredibly powerful event was hurtling toward us, only reaching us, finally, in 2008. By the same token, there are possibly even more spectacular cosmic explosions that have already occurred, but which we might not discover for thousands or even millions of years, because their light has not yet had time to reach us.

What all this means is that the sky is a time machine. The farther out you look into space, the further back you see into time. Want to see what the Universe looked like a billion years in the past? Easy—just find some stars and galaxies that are a billion light-years away, and you will see them not now, but how they appeared an incredibly long time ago.

This neat trick is the vital key we need to find the elusive stars of Population III. In our own Milky Way, all these stars are probably extinct. But if we look at very distant galaxies, it's possible that we might be able to look back in time and see some of the long-dead stars from Population III, their light just reaching us now after a journey of billions of light-years.

If we have a powerful enough telescope to see these extremely faint stars, and if we know where to look, we will see the very first stars in the Universe, formed not from the stardust produced by earlier generations, like our Sun and solar system, but constructed from the raw materials produced in the Big Bang itself. Like the discovery of early human fossils by paleontologists, the detection of the first stars in the Universe will give us dramatic new insights into our origins.

Live quietly, live forever

The Sun is about halfway through its life. As discussed in chapter 1, at an age of around 10 billion years (about 5 billion years from now), virtually all the hydrogen in the Sun's core will have been converted into helium. The Sun will move into the end stages of its life, before burning itself out and gently puffing off its outer layers to form a beautiful planetary nebula. SDSS J1029, with 13 billion years under its belt, is already older than the Sun will ever be. However, it seems quite healthy, and should have many years of life ahead of it. In contrast, there are other stars, like the bright star Betelgeuse in the constellation Orion, which are less than 10 million years old but are already well into old age.

While all these numbers seem equally enormous, there is in fact a vast contrast in the ages and lifetimes of different stars. To put this in perspective, let's suppose that the Sun is 40 years old, rather than 4.6 billion. In this case, the Sun will live to be about 80, a solid, respectable life span. In comparison, Betelgeuse is then only a four-week-old infant, but is already at the end of its life. And SDSS J1029 has just turned 100, but shows no signs of slowing down.

What is different about these three stars, that their life spans are so astonishingly different? It all comes down to a star's mass. The more massive a star, the hotter the fusion reactions at its core, and the faster it burns its fuel. And the rate of fuel consumption increases so dramatically for larger stars that even though they have more fuel to burn (because they are bigger), they still use it all up much faster, compared to low-mass stars. For the largest stars, the mantra "Live fast, die young" very much applies.

So it is no surprise to learn that Betelgeuse is a behemoth, weighing about 20 times the mass of our Sun. That factor of 20 in mass forces it to empty its fuel tank about 1,000 times faster than the Sun. In contrast, the mass of SDSS J1029 is only about 80% that of the Sun. That 20% difference, although seemingly modest, means that SDSS J1029 burns its fuel at a much more miserly rate than our Sun, providing it with greatly increased longevity.

Given this tight link between a star's mass and its lifetime, are there even smaller stars than SDSS J1029 that live for far longer? Indeed, the longest-lived stars in the Universe are tiny stars known as "red dwarfs," so named—you may not be surprised to hear—because of their size and their color. Red dwarfs are easily the most common type of star in the Milky Way, but are difficult to see because they are so dim. Indeed, 20 of the 30 nearest stars to the Sun (including the nearest of all, Proxima Centauri) are red dwarfs, but none of these 20 is bright enough to be seen with the naked eye. Red dwarfs are very much the silent majority!

Red dwarfs are typically 10–40% the mass of the Sun. The temperature at their cores is enough to keep nuclear fusion going only at the barest level, so they consume their hydrogen fuel unbelievably slowly. This alone is enough to lead to a very long life, but red dwarfs also have a very different structure from a star like the Sun that allows them to live even longer. In the Sun, the only hydrogen available as fuel is that in the core—the rest of the Sun never gets hot enough to undergo fusion, and so, even at the end of its life, most of the Sun's hydrogen will be untouched. In comparison, the gas inside a red dwarf is constantly being churned and stirred by swirling motions in the star's interior. This means that all the gas inside the star repeatedly drifts into the core and out again, allowing such stars to burn their entire mass, rather than just the core region. These extra fuel tanks allow red dwarfs to shine for unimaginable spans of time before they exhaust their gas supply, giving them lifetimes of around a trillion years. Returning to our analogy of a 40-year-old Sun, a typical red dwarf would then have a lifetime of 4,000 years!

For most stars, we can compare our theories of stellar evolution to actual observations of stars at various stages of their lives. But for red dwarfs we can only rely on our calculations, since all the red dwarfs we can see are still effectively in childhood. While we do not yet know what the first stars looked like, it seems likely that the last stars to shine before the Universe goes dark forever will be red dwarfs, which will keep smoldering feebly for eons long after all the other light in the cosmos has been exhausted.

Faster than a kitchen blender

Almost everything in astronomy happens on time scales much longer than a human lifetime, but occasionally, the Universe can surprise us.

Take supernovas, for example. As we have seen in chapters 1 and 2, these are the momentous explosions that mark the death of especially massive stars. Such stars have evolved steadily over several million years up to that point, but at the end, everything suddenly comes to a head quite rapidly, even by our standards. The collapse of the star's iron core that triggers the supernova only lasts a tiny fraction of a second. And the resulting shock wave takes just an hour or two to work its way outward through the star, until it reaches the surface and rips the outer layers of the star apart.

Usually, the entire star is not destroyed in the explosion. Left behind is what was once the iron core, heavier than the Sun, and now collapsed into a ball of neutrons just 15 miles across: a neutron star. Before the explosion, the star might have taken around 10 hours to perform a single rotation on its axis. But after the supernova, the remnant neutron star spins at a much, much more frenetic pace.

We know this because, for reasons we don't yet entirely understand, some neutron stars produce one or more strong narrow beams of radio waves that emanate from a fixed point on their surface. As such a star rotates, this beam sweeps across the sky like the light from a lighthouse. And if the Earth happens to lie in the path of this beam, we see the spinning neutron star as a "pulsar," a celestial radio beacon that appears to flash or pulse once per rotation. Every such pulsar flashes on and off at its own individual rate. But for a typical pulsar, the pulses are seen about once per second. This implies that each rotation takes a second to complete. That's 86,400 rotations per day, a far more rapid rotation rate than any ordinary star.

So far, almost 2,000 such pulsars have been identified, and in each case the spacing between the pulses can be measured extremely precisely, giving us incredibly accurate measurements of the rates of rotation of these tiny, distant stars. Careful study of individual pulsars over months and years shows that in almost all cases, these cosmic clocks are gradually slowing down, the times between each "tick" almost imperceptibly lengthening. We'll look into the mechanism behind this slowing down in chapter 8. But for now the important thing to understand is that because all pulsars are braking in their spin rates, these stars are expected to have been born spinning much faster than the spin rate we see for them today.

Confirming this idea is a bright supernova that was seen in the constellation of Taurus in July of the year 1054, and that was recorded by civilizations all over the world. In 1968, a pulsar was discovered at the exact position of this explosion, the direct identification of the neutron star left behind by the earlier supernova. With an age of less than a thousand years this pulsar is among the youngest known, and is also one of the fastest spinning, rotating 30 times per second. In 1998, an even faster young pulsar was discovered in the constellation of Dorado—this one, also only a couple of thousand years old, is spinning at an incredible 62 times per second. We do not know how fast either of these pulsars was spinning when they were first formed, but presumably it was even faster than we see them spinning now. A rough guess is that when pulsars are formed in supernova explosions, they are perhaps spinning at least 100 times per second.

If pulsars are born spinning very rapidly, and if they gradually brake in their rotation as they age, then it's natural to expect that millions of years after a supernova explodes, its corresponding pulsar will have slowed down substantially.

Indeed, some old pulsars spin only once every 5–10 seconds. This is still ridiculously rapid compared to most stars and planets, but is glacially slow for a pulsar.

This all seems relatively straightforward. But bizarrely, some pulsars, very late in their lives, can reverse this gradual slowdown, and despite now being hundreds of millions or even billions of years old, are seen spinning more rapidly than at any previous points in their lives. The current record holder is a pulsar in the constellation Sagittarius with the name "PSR J1748-2446ad," which was discovered in 2006 by a Canadian astronomy student, Jason Hessels. Hessels's extraordinary measurements showed that this particular star is spinning *716 times per second!* And what's more, this and the many other rapid rotators (there are more than 100 known pulsars spinning faster than 200 times per second) are not only spinning unusually fast, but are barely slowing down at all. A billion years from now, PSR J1748-2446ad will still probably be spinning more than 500 times per second. These ancient neutron stars seem not only to have somehow reversed the aging process, but to have discovered the secret of eternal youth.

What causes a pulsar, having spent most of its life slowing down, to later spin back up to rotation rates faster than it has ever experienced before? The typical locations of these fast-spinning pulsars offer a vital clue. In most parts of the Milky Way Galaxy, the vast majority of pulsars are relatively normal (in as much as a 15-mile ball of neutrons can ever be normal!), spinning about once per second, and gradually slowing down—the superfast rotators are very rare. However, pulsars are also seen in globular clusters, and here the situation is reversed—almost all pulsars in globular clusters are old neutron stars rotating ultra-quickly, and normal pulsars are virtually never seen.

As we saw in chapter 2, the very dense stellar environments of globular clusters can open up all sorts of strange and unlikely possibilities for stellar evolution, possibilities that would almost never occur under normal circumstances. And this is thought to be the case for old, fast-spinning pulsars. The currently accepted theory for how most of these strange beasts are produced is that an ordinary pulsar in a globular cluster passes very near to another star, and the two capture each other into a close orbit. If the orbit is sufficiently small, and if the other star is large enough, then the gravity of the pulsar will strip gas off the surface of its new companion, and drag it down toward its surface. As this gas swirls downward to the pulsar in a furious spiral, the orbiting gas can exert drag on the pulsar's rotation, gradually slowing it down or spinning it up, depending on the exact conditions of this bizarre process. Given sufficient time, this can make a pulsar spin faster and faster, until it reaches rotation rates of hundreds of times per second. For this reason these rapid rotators are known as "recycled" pulsars—having been reenergized by their companions.

Because recycled pulsars spin unimaginably fast, and also barely slow down with the passage of time, they make remarkable clocks. Over periods of years, these pulsars rival the most exquisite laboratory clocks on Earth for their accuracy and stability. While we don't have any plans to redefine Greenwich Mean Time using pulsars, they do allow us to perform amazing calculations and measurements on celestial orbits.

To see what I mean, first consider the orbit of the Earth

around the Sun. We all know that the Earth takes a year to revolve around the Sun, with one year being 365 days, 5 hours, 48 minutes, and 45 seconds (the extra 5 and a bit hours is why we need leap years: 365 days aren't quite a year, so we need to add an extra day every 4 years to keep things in sync). The average distance from the Sun to the Earth is 92.96 million miles, but the Earth's orbit isn't perfectly circular. Instead, it takes the shape of an ellipse or oval. Because of this, the exact Earth to Sun distance changes throughout the year. Specifically, at the start of January each year, the Earth is at its closest distance to the Sun, about 1.6 million miles less than the average. And near the start of July, the Earth reaches its farthest point from the Sun, 1.6 million miles more than the average. So over the course of a year, the radius of the Earth's orbit stretches and shrinks by up to 3.2 million miles.

With this in mind, let's now consider an extraordinary pulsar known as "PSR J1909-3744," which is 3,700 light-years away in the constellation Corona Australis. PSR J1909-3744 is a recycled pulsar in a binary orbit, and endlessly circles its companion star. But because PSR J1909-3744 is also rotating 340 times per second, we can use it as a super-accurate clock to make some extraordinarily precise calculations.

For example, even though the pulsar is several thousand light-years away, we can determine that the time taken by the pulsar to complete one orbit is 36 hours, 48 minutes, and 10.032524 seconds (with an accuracy of about one microsecond!). What's more, the orbit that this pulsar traces is the most perfect circle known in the Universe. The diameter of the pulsar's orbit is 708,000 miles, but instead of shrinking in and out by a sizable amount over the course of one circuit, as is the case for the Earth, over 36 hours the size of the pulsar's orbit changes by around a thousandth of an inch, less than the width of a human hair. That we can make such exquisite measurements on a tiny object so far away is nothing short of astonishing.

And if that isn't enough, the spin rates of recycled pulsars could also help us understand exactly what the atomic structure of neutron stars might look like. As we will see in chapter 10, we think that the incredible density of a neutron star squashes atoms into strange new shapes, resembling long tubes and flat pancakes. But the full details of how atoms behave and shape themselves under these extreme conditions remain only an educated guess, because we are not in the position to be able to travel to a neutron star, scoop up a lump of material from its surface, and bring it back to the lab for analysis!

Helpfully though spin rates may soon provide a crucial clue. That's because there is a firm physical limit to the fastest possible spin rate that a pulsar can achieve. Spinning an object subjects it to an apparent centrifugal force, in which everything inside it feels pushed toward the outside. Spin it faster than its structure can handle, and the object flies apart. Conversely, the faster we see an object spinning, the stronger its structure must be. We can therefore take each possible model for the structure of a neutron star, and calculate the fastest spin rate that such a structure can withstand. If we find a neutron star spinning faster than this limit, then we immediately know that the corresponding model for neutron star structure is probably not correct.

PSR J1748-2446ad, spinning at 716 times per second, is able to eliminate some models of neutron star structure, but several others can still easily hold the star together, even at this furious rate of rotation. The race is thus on to find even faster spinning pulsars, that will push the different theories to their limits. Astronomers often talk reverently about the quest to find a "sub-millisecond pulsar"—that is, a star spinning more than 1,000 times per second. Do such cosmic hummingbirds exist? Time will tell.

4

EXTREMES
OF·SIZE

Looking up at the night sky, it's easy to understand why for many centuries most people thought the Earth was the center of the Universe. After all, our planet is big, while the Sun, Moon, planets, and stars all look pretty small. Even today, the Earth still seems huge. We might be able to skim the world's cities with the flick of a computer mouse, thanks to applications like Google Earth, but travel any appreciable distance and the full 200 million square miles of the Earth's surface begin to become apparent.

I sometimes drive from Sydney to Melbourne, a distance of almost 600 miles. This journey takes an entire day, and I inevitably clamber out of my car at the end of the journey feeling exhausted and drained by the experience. And yet looking at a globe, I feel cheated and disappointed—I feel as if I've traveled so far, and yet Sydney and Melbourne sit almost on top of each other when viewed in the context of a planet that's almost 8,000 miles in diameter.

Incredibly, modern airplanes allow us to travel from one

side of the planet to the other in less than 24 hours. This might greatly shorten the time taken for the journey, but I don't think it makes the world seem any smaller. I've often sat by the window of a plane on a clear day, and looked at the landscape spread out before me. What never fails to amaze me on these trips is that although I don't ever doubt that the Earth is round, the evidence of my own eyes is that the world seems utterly flat, even when viewed from 30,000 feet. I stare and squint at the horizon, perhaps hundreds of miles away, and even though I know the Earth is curved, I just can't see it. I have no real interest in ever going into space, but I think I'd like to be taken up 20 or 30 miles, just for a moment, so I could see for myself that the world really is round.

And of course what we think of as the Earth is merely the surface of a three-dimensional sphere. The full volume of our planet's interior is almost 300 billion cubic miles, almost all of which is completely unexplored. So by our everyday standards, our planet is enormous. In contrast, the Sun and Moon seem small enough that they can be completely hidden by an outstretched hand, while all the planets and stars are mere pinpricks, seemingly having no size at all. Looking at the night sky, one can start to appreciate why the gradual revelations that the Sun, not the Earth, is the dominant member of our solar system; that the Earth isn't even the largest of all the Sun's planets; and that the Sun itself is only one of many stars were all such seismic shocks to the sensibilities of the Church and of the scientific community.

Even though we now accept that the Earth and Sun are small and unimportant on a cosmic scale, it's hard to truly appreciate the vast sizes of these familiar entities. After all, the

Sun is 860,000 miles across, big enough to fit a million Earths inside it and still have plenty of room to spare.

These numbers are already beyond our comprehension, and yet the Universe contains many objects that are much, much larger.

Bloated, pulsing, and leaking

The Sun is an ordinary and unremarkable star in almost every respect. Its age, composition, temperature, and mass are all quite typical. And while there are many stars much smaller than the Sun, some stars are vastly larger.

Some stars are bigger than the Sun simply because they have a greater mass. For example, the bright star Achernar, visible to the naked eye, has about six times the mass of the Sun, and so is 10 times larger, with a diameter of around 9 million miles.

Achernar is broadly similar to the Sun in that it is fusing hydrogen at its core to form helium. As we saw in chapter 1, stars generating their heat in this way are said by astronomers to be in the main sequence phase of their lives. And indeed if a list of main sequence stars is ranked by increasing mass, these stars will also form a steady sequence of increasing size (as well as a sequence of drastically decreasing lifetime, as I explained in chapter 3).

However, there are many stars in the sky that deviate wildly from this simple ordering. One famous example is Mira, a star about 420 light-years away in the constellation Cetus. Mira is only about 20% heavier than the Sun, but is 400 times larger! The diameter of Mira is around 300 million miles—so big that were we to replace the Sun with Mira at the center of the solar system, Mercury, Venus, Earth, and Mars would all sit inside it. Even from the distance of Neptune, Mira would seem enormous, appearing 180 times larger in the sky than the Sun does from Earth.

Mira's enormous size presents a puzzle. If this star is only slightly more massive than the Sun, why is it larger than the Sun by such a stupendous factor?

A vital clue is provided by the name Mira itself, given to this star by the 17th-century Polish astronomer, Johan Hevelius. *Mira* is Latin for "wondrous" or "extraordinary." Hevelius chose this name because this star does something that, at the time, was unique: Sometimes it disappears.

In our modern understanding, Mira is a "periodic variable star," meaning that it changes in brightness in a predictable, regular cycle. In Mira's case, this cycle lasts 11 months. At the start of this cycle, Mira is a reasonably bright star, easy to observe with the naked eye. But it then slowly fades. Five months later it vanishes from view, and can then only be seen with binoculars or a small telescope. Another two months after that, Mira reaches the faintest point of its cycle, having dropped to a brightness hundreds of times below its peak. Over the next four months, Mira then brightens again, returning to naked-eye visibility before the process begins again. We have records of Mira's gradual 11-month cycle going back almost 400 years. What's more, we have now identified and cataloged several thousand other stars that similarly gradually brighten and fade, over times ranging from 10 weeks to three years. In honor of the first known of their kind, this group of stars is known as "Mira variables."

Mira's enormous size and regular changes in brightness are both signs that it is moving into old age, and is no longer in the best of health. Mira was once a normal main sequence star like the Sun, but has now consumed all the hydrogen in its core, leaving helium as the residue of its billions of years of furious energy production. A star that reaches this point—its energy supply exhausted—no longer has radiation and heat to support its bulk, and begins to be squeezed by the relentless fist of gravity. The star's helium core contracts, and the extra heat that this puts out then ignites nuclear fusion in a surrounding shell of hydrogen.

The star has a new energy source, saving it from extinction, but it pays a heavy price, because the combined heat of the core and of the new shell of burning hydrogen causes the outer layers of the star to dramatically expand. The star becomes hundreds of times larger, thousands of times brighter, but with about half the surface temperature it had in its youth. The result is a star with a dense, hot core surrounded by an enormously bloated, tenuous envelope, now shining a dull red rather than the bright yellow we associate with a star like the Sun. The star has now become a "red giant."

Although many stars (including our own) will become red giants, these are not especially common stars, because this is a relatively short period of a star's life. For the Sun, the red giant phase will last for about a billion years, an unimaginably long period by everyday standards, but brief compared to the 10 billion years of the main sequence that will have preceded it. Despite their rarity, red giants are so luminous that many are easily seen with the naked eye, even if they are hundreds of light-years away. Apart from Mira, other well-known red giants include Arcturus (the third-brightest star in the sky) and Aldebaran (the brightest star in the constellation of Taurus).

In contrast to Mira's dramatic changes in brightness, Arcturus and Aldebaran shine reasonably steadily. Even among red giants, Mira is relatively unusual. This is because this is not Mira's first turn at being a red giant. As I explained in chapter 1, the inert helium core of a red giant will eventually be heated by its own crushing gravity to a temperature of around 180,000,000°F, allowing fusion to begin again at the center. At the center of the star, helium is now being converted into carbon, surrounded by a shell of hydrogen burning into helium. The star has now begun the "horizontal branch" phase of its life, a period of relative calm lasting another hundred million years or so. Comparatively normal operations now resume, and the star contracts to a more manageable size.

But when the helium at the core has all been converted into carbon, the star finds itself in a similar situation to when the hydrogen in the center had all been turned into helium. Nuclear reactions in the core cease, the core contracts and heats up, and this then ignites a surrounding shell of helium. We now have a star, billions of years old, that has a carbon core, a layer enveloping this in which helium is burning to make carbon, and another layer farther out in which hydrogen is burning to make helium. Once again, this extra heat makes the star brighten dramatically and swell to enormous size, with a diameter larger than the orbit of the Earth. The star resumes its red giant persona; astronomers refer to stars going through this for the second time as being on the "asymptotic giant branch," or "AGB" for short. And this is where Mira and its brethren, the Mira variables, now find themselves.

Mira varies so dramatically and regularly in its brightness because, like a slowly beating heart, it is pulsating in and out. When Mira is at its brightest, it is about 20% smaller than when it is at its faintest. These pulsations are symptomatic of a fundamental breakdown in the rules that usually govern a star's behavior.

Luckily for us, the Sun and most other stars are finely tuned engines, with everything in perfect balance. Should the Sun start to get a little hotter than usual, the increase in temperature will cause the gas in the interior to become a little more transparent, allowing this extra heat out, and bringing the temperature back down to the desired level. Similarly, if things cool down, the gas will become more opaque, holding the heat in. This wonderful naturally occurring thermostat keeps most stars steady in their brightness, temperature, and size.

A symptom of Mira's advanced years is that its thermostat is no longer working correctly—in fact, its behavior has turned completely upside down. The gas inside Mira happens to be at a density and pressure where its properties become rather unusual. When a layer of gas inside Mira heats up, the gas becomes slightly more opaque, rather than slightly more transparent as usually happens in other stars. This traps the heat inside the star, building up the pressure until the star is forced to expand. As the already enormous star bloats up to an even larger extent, the gas is able to cool down. Next, instead of the gas becoming more opaque to hold the heat in, it becomes transparent, so that the heat leaks out even faster, causing the star to fall back down to its earlier size. As Mira now shrinks, the pressure builds up, the temperature rises, and the cycle begins again.

Another symptom of old age is that Mira is leaking. While the star pulses in and out, at the same time about 20 trillion tons of material are streaming every second from its surface into space. Mira happens to be moving quite rapidly through space, at a speed of about 290,000 miles per hour: The combination of a high speed and rapid weight loss has left an incredible trail behind the star, extending over 13 light-years (about four times the diameter of the full Moon). As Hevelius first noted almost 400 years ago, Mira truly is wondrous.

The stellar heavyweights

Incredibly, there are stars much larger even than Mira. A star weighing more than about 10 times the mass of the Sun hits the same problem as Mira when its core runs out of fuel: Nuclear reactions shut off, the core is squeezed by its own gravity, and hydrogen begins burning in a shell around the core. But the larger mass means more gravity, a tighter squeezing, and a higher temperature: The star swells up to a much larger size than a star like the Sun, becoming a red supergiant.

For the brightest of these behemoths, the red hue is easily distinguished even to the naked eye. Prominent examples include Antares in the constellation Scorpius, and Betelgeuse in Orion. These stars are absolutely gargantuan: Antares has a diameter 800 times that of the Sun, while Betelgeuse is even larger, with an extent of almost a billion miles, more than 1,000 times larger than the Sun.

But the freak show doesn't stop there. Astronomers have now identified a class of stars that they have termed "hypergiants"—stars of extreme mass that correspondingly expand to extreme sizes when they exhaust their central hydrogen supply and leave the main sequence. Hypergiants are rare and hard to find, but their vast girth more than makes up for their small numbers.

The largest known star is a hypergiant known as "WOH G64," at a distance of about 160,000 light-years in the constellation Dorado. This extraordinary star has a diameter of around 1.2 billion miles, 1,500 times that of the Sun. Placed at the center of our solar system, WOH G64 would envelop not just Mercury, Venus, Earth, and Mars, but would comfortably extend far past the orbit of Jupiter as well. As a comparison, if WOH G64 were the size of a basketball, the Sun would be smaller than a dust mite. And the Earth? Our home planet, which even from a plane window seems to stretch on forever, would be reduced to the size of a bacterium.

Let's stick together

Even the largest stars are just tiny members of the vast stellar chorus that makes up an entire galaxy. Our Milky Way is a typical example of a "spiral" galaxy: a flat disc of several hundred billion stars, many grouped into spectacular, curved, spiral arms. This glittering structure spans 100,000 light-years from one side to the other, and slowly rotates around its center, taking about 200 million years to complete each revolution. Since the Earth was first formed 4.6 billion years ago, the Milky Way has spun around only 22 times.

The Milky Way is incredibly ponderous, with a mass more than that of a trillion Suns. It thus exerts a strong gravitational force on its surroundings, and can easily ensnare smaller galaxies that wander too close. Indeed, astronomers have identified many stars in our Milky Way that have unusual compositions or strange orbits—these are thought to be the remnants of galaxies that the Milky Way captured and then ultimately digested!

In the case of the Sagittarius Dwarf Elliptical Galaxy, or "SagDEG," the process of digestion is taking place as we watch. SagDEG was only discovered about 15 years ago, because what's left of this galaxy is mixed in with the much larger number of stars from the Milky Way. SagDEG consists of only a few tens of million of stars; the rest have been gradually stripped away and are now spread all over the Milky Way.

The Milky Way barely skips a beat as it swallows smaller galaxies like SagDEG. But the neighborhood bully will finally meet its match when it eventually collides with another big spiral, the Andromeda Galaxy. As I'll discuss further in chapter 9, the Milky Way and Andromeda are already in each other's gravitational thrall, and are hurtling toward each other at 270,000 miles per hour. But since the two galaxies are currently separated by about 2.5 million light-years, we will need to wait around 2 billion years before this colossal encounter reaches its climax.

The collision with Andromeda will be very different from those that the Milky Way has undergone with various smaller galaxies. Since Andromeda and the Milky Way are about the same size and mass, there can be no clear victor. During a complicated and chaotic dance lasting billions of years, long trails of stellar and gaseous debris will be torn away from each galaxy by the gravitational might of its partner. But despite this large-scale carnage, the odds of individual stars colliding are small.

Eventually the two delicate spiral galaxies will settle into a single giant ball of stars, known as an "elliptical" galaxy. Astronomers have identified many elliptical galaxies, most of which are thought to be the results of mergers of smaller spirals. A newly formed elliptical galaxy will have more mass and more gravity than either of the galaxies that formed it, and so will be even more enthusiastic about getting to know its neighbors. As elliptical galaxies progressively absorb other galaxies, they can become very large indeed.

Taking this process to its extreme, the largest galaxies in the Universe (known, for obscure reasons, as "cD galaxies") are perhaps the product of dozens of mergers. The largest of these gargantuan beasts is IC 1101, a galaxy an incredible 5 million light-years across (about 50 times the size of the Milky Way), and containing many trillions of stars. To put this in perspective, let's now shrink WOH G64, the largest known star, down to the head of a pin. The Milky Way would then be around 200 miles across, while IC 1101 would be the size of the entire planet Earth.

Voids, bubbles, and walls

Galaxies, regardless of their size, are mere flecks on the Universe's canvas. The galaxies we see in stunningly detailed photographs, filled with whorls of color and glorious swaths of stars, are the exception rather than the rule. Most galaxies are at much larger distances, and even through powerful telescopes appear as little more than elongated blobs.

And yet, despite offering so little information individually, these distant galaxies are often of far more interest to astronomers than the closer ones that we can study in detail. There are huge efforts taking place to conduct forensic experiments on the entire Universe: to study galaxies, not for their own sake, but to use them as buoys on the cosmic sea.

If one sifts through the enormous catalogs of galaxies now available, the initial impression is that galaxies are scattered randomly and evenly throughout the cosmos. However, the reality is very different.

The problem is a lack of perspective. When we look at the everyday world around us, we can discern which objects are nearby and which are farther away through the depth perception that we get from having two views of the same field from slightly different angles, one from each eye. However, this trick doesn't work when looking at the night sky, because everything is so far away that the view through one eye is no different from that through the other. The vast distance scale of the cosmos makes us all into one-eyed stargazers, and the sky appears flat.

Over the last few decades, astronomers have tried to move beyond the single smooth surface that the sky usually presents, and to reconstruct a true picture of the Universe. The goal has been to discover very large numbers of galaxies, to measure their positions, and, crucially, to also determine their distances. Cast into three dimensions, the sky is then transformed: What before looked like a random sprinkling of salt grains on black paper becomes a remarkably complicated and textured sculpture. What we now know is that the Universe is not smooth, but is frothy and lumpy. Galaxies are clustered into long filaments and chains, joining up to other groups of galaxies like interlocking soap bubbles. Conversely, inside these bubbles are vast empty voids: dark expanses of space, free of any stars or galaxies, often more than 100 million light-years across (a lot more on voids in chapter 10).

These three-dimensional reconstructions have revealed the largest known structure in the Universe, a colossal filament of thousands of galaxies known as the "Sloan Great Wall," discovered by American astronomer J. Richard Gott and his colleagues in 2003. The Sloan Great Wall is approximately 1.4 billion light-years across, and runs behind the constellations of Hydra, Sextans, Leo, and Virgo, stretching across almost a quarter of the sky. It is not a single linear thread, but writhes and twists, even splitting up into two separate tendrils for a few hundred million light-years, which then rejoin farther along.

It's remarkable that something as enormous as the Sloan Great Wall was not discovered until 2003. You would think that all we would need to do is photograph the night sky, and we would see such a feature as a giant chain of galaxies, spanning the heavens. But while we can easily see the individual galaxies that make up the Sloan Great Wall, without the perception of depth the shape of the wall is completely washed out. The many more unrelated galaxies in the foreground and background hide the Sloan Great Wall from view.

Measuring distances to galaxies is hard work. It took decades of heroic effort to get to the point where we had a good enough three-dimensional picture of the Universe to be able to discern a structure like the Sloan Great Wall. The way distances are measured in these experiments is to split up a galaxy's light into a spectrum of that light's constituent wavelengths, and to thus very precisely determine the galaxy's color.

Because the entire Universe is expanding, the light waves from distant galaxies are stretched as the light travels toward us. This stretching of the light waves causes these galaxies to appear redder than they would be if they were nearby. The farther away a galaxy is, the greater this shift to the red end of the spectrum. So if we can measure the "red shift" of a galaxy, we can then make a reasonable estimate of its distance. For a typical galaxy, using a modern telescope, such a measurement might require an exposure of an hour.

So why are these measurements hard work? Because to build up a three-dimensional picture of the Universe requires repeating this process for hundreds of thousands of galaxies. What takes just an hour for one galaxy becomes a decade or more if one sets out to cover enough galaxies to make a useful picture. The early, pioneering efforts in the 1980s indeed approached this task one galaxy at a time, and produced the first pictures of bubbles, walls, and voids. But in the 1990s, astronomers realized that new technology was needed to cover significant parts of the sky, and so they designed new cameras that use hundreds of optical fibers to measure the colors of many galaxies at once. With these fiber systems, a single telescope can be effectively converted into a hundred separate telescopes, all simultaneously measuring distances to different galaxies.

These new multi-fiber surveys have now covered about 30% of the sky, leading to the discovery of the Sloan Great Wall, plus many other remarkable structures. As stunningly successful as these telescopes have been, it is sobering to realize that 70% of the sky is yet to be mapped. As we continue to unveil the true shape of the cosmos, it remains to be seen whether something even bigger than the Sloan Great Wall will emerge.

The runt of the solar system

Lest I leave you feeling overwhelmed at the stupendous size of the largest objects in the Universe, let's finish with some of the smallest objects we have discovered.

Of course the very smallest things are not seen through our telescopes, but through powerful microscopes: molecules, atoms, protons, neutrons, and quarks. And modern physics predicts a variety of particles that have no size at all, including the humble electron that provides us with electricity (see chapter 8). Attempts to measure the size of an electron have left us only with upper limits: If this tiny fleck of matter does have a size, it must be less than 0.0000000000000000004% of an inch.

Our studies of the heavens cannot come close to probing such tiny scales, but there are certainly many objects we know of that are surprisingly small, given the large distances we must peer across to see them.

As a benchmark, let's start with the Sun, which is 860,000 miles across. As we saw in chapter 3, the smallest normal stars are red dwarfs, which can be 10% or less than the mass of the Sun, and correspondingly can be as small as 120,000 miles across.

But once stars reach the end of their active lives, they can get far smaller. When the Sun exhausts its fuel, the core will be left behind as a white dwarf. White dwarfs are hot (chapter 1), dense (chapter 10), and are only about 6,000–7,000 miles across, roughly the same size as the planet Earth. A star much larger than the Sun explodes in a supernova, leaving behind an even denser, smaller object than a white dwarf, a neutron star. Neutron stars are thought to have diameters of only 15 miles, smaller than most cities.

If we restrict ourselves to studying objects hundreds or thousands of light-years away, it is challenging to see anything smaller than a neutron star. Even the smallest planet known outside the solar system, a tiny object called "PSR 1257+12A," is well over 500 miles across. But if we venture closer to home, into the realm of moons, comets, and asteroids orbiting our own Sun, we can quickly find even smaller objects.

We usually think of the solar system as the Sun and its eight planets. However, this is just the tip of the iceberg. Also circling the Sun are five "dwarf planets" (a new category that includes the ex-planet, Pluto, plus four other similarly sized objects), approximately 350 moons, around 4,000 known comets, and more than 500,000 known asteroids.

If you buy a complicated model kit made out of Lego or Meccano, you will often find that there are a few pieces left over when you've finished—the box has more pieces than you actually needed. In the same way, asteroids are the "spare parts" of the solar system. When the Sun formed, most of the boulders, rocks, and pebbles in orbit around it eventually coalesced into the planets and moons that we see today. But for various reasons some of this debris did not form planets or moons, and remained in more or less their original condition to become the asteroids.

More than 15,000 of the asteroids have names, ranging from the classical ("Juno," "Vesta," and "Urania"), to the reverential

("Einstein," "Beethoven," "Hitchcock," and "DiMaggio"), to the methodical ("United Nations," "Chicago," and "NASA"). But the vast majority simply have an alphanumeric code, along with a catalog entry containing the details of their orbit, size, and mass.

Most known asteroids range in size from a few hundred miles down to a few hundred feet. There are uncounted millions of objects much smaller than this, but they are either too faint and far away to be seen, or stray so close to us that they burn up in the Earth's atmosphere before we know that they were even there. But there is now a growing number of truly tiny objects that pass close enough for us to see and track them as they shuttle their way around the Sun.

The current record holder for the smallest known object in orbit around the Sun is a tiny runt of an asteroid known as "2008 TS26." It was discovered on October 9, 2008, by renowned asteroid and cosmic hunter Andrea Boattini, using a small telescope near Tucson, Arizona. Boattini saw a tiny speck of light, about a million times fainter than possible to see with the naked eye, in the constellation of Pisces. Over the course of the next couple of hours, this little object drifted across the sky by more than the diameter of the full Moon, clearly giving away that it was no distant star or galaxy, but had to be something much closer to home.

With a few more observations Boattini and others were able to calculate the basic properties of 2008 TS26, and found that this lump of rock was absolutely minuscule. How small? Asteroid 2008 TS26 is only about 2–3 feet across, the size of a large beach ball! It's a superb testament to both the quality of modern instruments and the precision of our calculations that astronomers can find and track such tiny objects.

Interestingly, the orbit of 2008 TS26 reveals that just a few hours before it was discovered, it had almost collided with the Earth—it missed us by just 4,000 miles before heading back out into space. While this was one of the closest shaves ever recorded, it was not great cause for alarm. Had 2008 TS26 scored a direct hit, its small size meant that it would have largely disintegrated in the Earth's atmosphere, with only fragments reaching the ground. Nevertheless, with an orbit around the Sun that takes only 32 months, little 2008 TS26 will no doubt be returning to our neighborhood in the near future to have another try.

5

EXTREMES
OF SPEED

Every four years the world turns to the Olympic Games and marvels at the dazzling speed of the competitors in the 100-meter sprint. Finely tuned athletes cover the distance in about the same time it takes most of us to tie our shoelaces. The fastest of these runners can hit speeds of around 25 miles per hour for a furious few seconds as they hurtle for the finish line.

But put in context, even these amazing sprinters are not exactly fast movers. With the help of technology, we can reach much higher velocities. The greatest speed most of us ever attain is in an airplane, where a typical cruising speed is around 550 miles per hour. The land speed record, set in 1997 by Briton Andy Green in a rocket-propelled car, is 760 miles per hour.

Astronauts and cosmonauts orbiting the Earth on the International Space Station easily exceed this, hurtling around the globe at around 17,000 miles per hour. And the fastest any human has ever traveled is 25,000 miles per hour, or almost seven miles every second, a speed hit by the American astronauts aboard *Apollo 10* during their return from the Moon in May 1969.

Impressive as these speeds seem, our efforts are far out-paced by even the most mundane astronomical objects.

Let's start with our own planet Earth, which orbits the Sun once each year. In order to complete this orbit, the Earth must maintain an average speed of more than 66,000 miles per hour. Even though we are ferried along at this frantic speed at every moment of our lives, we don't feel its effects, because the Earth is very nearly moving in a straight line: The curve of the Earth's orbit is very slight, changing course by only about one degree per day. So despite the high velocity of our orbit, this orbit is so big that the effect is virtually unnoticeable.

The Earth's speed in its orbit is far beyond the speeds we are used to experiencing in our everyday lives. But the Earth in turn is dwarfed by the frenzied speeds that many other celes-tial bodies routinely experience.

The fast and the furious

The Earth orbits the Sun at 66,000 miles per hour, but is easily outstripped by Mercury, the innermost planet in our solar system. Appropriately named for the fast-moving messenger to the gods, this small, hot rock orbits the Sun at more than 105,000 miles per hour.

Until the 1990s, Mercury was the fastest-moving planet we knew of. However, a flurry of new discoveries of planets out-side our solar system means that Mercury no longer ranks even in the top 100, and by modern-day standards is quite pedes-trian.

At the time of this writing, we know of more than 700 other planets orbiting other stars, with more being added to the list on a weekly basis. These other worlds, known as "extrasolar planets" or "exoplanets," are a strange group, usually bearing little resemblance to the familiar planets of our own solar system. The most remarkable discovery has been a category of planets known as "hot Jupiters." These are so named because they are huge gas giant planets like our own Jupiter, but which orbit extraordinarily close to their parent stars. While fast-moving Mercury sits at about 40% of the Earth's distance to the Sun and takes 88 days to carry out one orbit, a typical hot Jupiter is separated from its parent star by only about 5% of the Earth to Sun distance, and races around its entire orbit in just a couple of days.

One of the fastest-moving exoplanets currently known is a hot Jupiter called "WASP-12b." WASP-12b orbits an otherwise unremarkable star known as "2MASS J06303279+2940202" (or "2MASS J0630" for short), located 870 light-years from Earth in the constellation Auriga.

("WASP" stands for "Wide Angle Search for Planets," "12" because this was the 12th star for which the WASP project discovered an orbiting planet, and "b" to indicate that the planet is the second object known in this star system. "WASP-12a" would be the parent star itself, while if a second planet were discovered around this star it would be called "WASP-12c.")

WASP-12b is far too faint to be seen directly through any telescope—its parent star outshines it by a factor of 3,000. Instead, it was discovered through a simple but painstaking technique known as the "transit method."

Searching for planets orbiting other stars is a difficult

business. At the enormous distances we are talking about, planets are very small and extremely faint. To find a planet, it's much easier to adopt an indirect approach, and focus our efforts on a much bigger and brighter object—the parent star. The transit method is one of the techniques commonly used by astronomers, and relies on the fact that the orbits of at least some exoplanets will take them directly in front of their star from our viewpoint (in the same way that the Moon sometimes passes in front of the Sun, producing a solar eclipse).

While an exoplanet will not be anywhere near large enough to completely eclipse its star, it will block out some small part of the star, causing a very slight dimming in the light we receive from Earth. Of course this dimming only lasts for that small part of the orbit in which the exoplanet moves in front of the star. For the rest of the orbit, even a careful study will reveal no trace of the planet at all. And the vast majority of exoplanets will not have an edge-on orbit at all, and will never produce any such eclipse.

Finding exoplanets through the transit technique thus requires continuous, careful observations of very large numbers of stars, in the hope that just one of them will dim by a tiny amount for only a short period of time. Despite the effort involved, transit searches can be very fruitful, because the amount by which the star dims immediately tells us the size of the exoplanet, while the time between consecutive transits is the time it takes the planet to complete one orbit.

WASP-12b was discovered by a large group of astronomers led by Leslie Hebb at the University of St. Andrews in Scotland. Hebb and her team measured the brightness of 2MASS J0630 thousands of times throughout 2006 and 2007, eventually identifying a tiny dimming of the star's light by about 1.3%, lasting about 3 hours and repeating every 26 hours. We can immediately conclude that WASP-12b blocks out 1.3% of the light from its parent star during its 3-hour transit, and that the total time for one orbit around its star is 26 hours. This information, when combined with other measurements and calculations, tells us that WASP-12b is almost twice the diameter of Jupiter and that it hits a top speed in its orbit of 528,000 miles per hour, or about 150 miles every second!

Suppose we could somehow travel the 870 light-years that separates us from WASP-12b, and that we decided to dispatch an expedition to study it. Since WASP-12b is a gas giant, we could not land on its surface, but we perhaps might want to orbit the planet and study it from above. However, establishing such an orbit would not be an easy task, nor would we want to stay there for very long.

First, we would need to match speeds with WASP-12b before we could set up an orbit around it. This means similarly attaining a breakneck speed of 528,000 miles per hour around the parent star, just to keep up. This is more than 20 times the speed record held by *Apollo 10*, and so would require a combination of new propulsion technologies plus enormous amounts of fuel. (Then again, if we had already figured out how to travel from Earth to WASP-12b, perhaps this would not be a problem.)

If the WASP-12b mission planned to transmit its findings and images back to an audience on Earth or elsewhere, maintaining a communications link would take some extra effort. As we know from hearing an ambulance siren change in pitch as it passes us on the street, the sound from a moving object gets shifted to higher frequencies if it is traveling toward us, and to lower frequencies if it is moving away from us. This phenomenon, known as the "Doppler effect," also applies to radio transmissions from a distant space mission. As WASP-12b swings around its star, there will be a part of its orbit where it is moving at 528,000 miles per hour toward Earth; about 13 hours later, it will be moving 528,000 miles per hour away from us. The signal transmitted by anyone orbiting WASP-12b will thus experience huge Doppler shifts. Any radio antennas tracking this signal would need to continuously tune up and down the dial in order to maintain a lock.

And even if we could establish an orbit around WASP-12b, it would be an unpleasant experience. WASP-12b orbits its star at a distance of just 2.1 million miles. This sounds like a big number, but it's only 2.3% of the distance between the Earth and the Sun. At such close quarters, 2MASS J0630 would not be a warm, friendly beacon like the Sun, but would be a fiery beast dominating the sky, appearing about 5,000 times larger than the Sun does from Earth. Our intrepid explorers would need to heavily shield themselves from the bombardment of harmful X-rays and energetic particles. And to top it all off, at this close distance, the star's light and heat would be 6,000 times brighter than what Earth receives from the Sun. The resulting temperature would be extreme, around 4000°F. At this temperature, iron and gold would melt, while lead would boil. Our spacecraft would need to be made of extraordinary materials, and would need an equally extraordinary cooling and air-conditioning system to stop its human inhabitants from being incinerated. So far, scientists have found more than 100 hot Jupiters like WASP-12b, but this number is about to increase rapidly as a result of the increasingly sensitive and sophisticated methods for detecting exoplanets now at our disposal. This means that the planetary speed record currently held by WASP-12b will almost certainly be broken soon.

One of the questions that astronomers still puzzle over as they study hot Jupiters is how these bizarre worlds formed. It is unlikely that such large planets began their lives so close to their parent stars—the extreme gravity and temperature that they experience in their orbits mean their lifetimes are limited. Indeed, Hubble Space Telescope observations of WASP-12b suggest that large globs of gas are already beginning to be torn away from WASP-12b by 2MASS J0630. In another 10 million years or so, the planet may be completely shredded to pieces by its much bigger parent.

The explanation for hot Jupiters that most astronomers currently favor is that these planets formed far from their stars, in stately orbits similar to that occupied by Jupiter in our own solar system. But early in a star's life, the orbits around it are filled with a huge collection of gas, rocks, and dust. Through a process known as "planetary migration," Jupiter-like planets can then have their orbits slowly shifted by gravitational interactions with this interplanetary debris, gradually bringing them closer and closer to the central star. So perhaps the question to ask is not why other stars have hot Jupiters, but why our own solar system does not. As astronomers continue to discover more and more exoplanets, we move steadily closer to a full understanding of how planets form, and why different solar systems have such different distributions of orbits.

Run out of town

When we measure the speed of an object, the answer depends on our point of view. Standing by the side of the highway, cars seem to whiz past. But driving at the speed limit in the middle lane, cars in the faster lane seem to slowly overtake us, while those in the slow lane gradually drop behind. In the same way, any claims of a cosmic speed record need to be qualified by the vantage point from which they were measured. The ferocious speed at which WASP-12b orbits its star is measured from the perspective of the star itself. But what if the star is also moving through space at some high speed?

Indeed, from an external viewpoint, even our own Sun is a star in a hurry. The Sun is in an orbit around the center of the Milky Way Galaxy, as is every other star we can normally see with the naked eye. The Milky Way rotates very slowly—our solar system takes more than 200 million years to complete one loop. You might thus think that the Sun crawls at a snail's pace through the Galaxy. But the length of one circuit is a staggering 170,000 light-years, so even though we have 200 million years to cover this distance, there is no time for dawdling! Indeed, recent measurements have clocked the Sun's orbital speed around the Milky Way at around 568,000 miles per hour, slightly shading the speed of WASP-12b that we discussed above.

Although the Sun's speed sounds extreme, most of the stars in our neighborhood are following similar orbits, so usually we barely notice our motion at all. The orbits of many stars throughout the Milky Way have now been measured, and most seem to be orbiting at similar speeds to the Sun, similarly following gradual, circular orbits around the center of the Milky Way.

However, in the last few years, astronomers have discovered a very strange population of "hypervelocity" stars, which move much faster than stars like the Sun, and follow very different trajectories.

The story begins in 1951, when Australian astronomers Jack Piddington and Harry Minnett built a radio telescope in the southwestern suburbs of Sydney. Among their discoveries was an intense source of radio emission coming from the center of the Milky Way, in the constellation Sagittarius.

This object, now known as "Sagittarius A," has been the subject of intense study over the last 60 years. We now know that it is indeed located at the center of the Milky Way, approximately 27,000 light-years from Earth. Embedded within Sagittarius A is a tiny faint object known as "Sagittarius A*" (pronounced "Sagittarius A-star"), which as far as we can tell sits at the exact geometric center of our Galaxy. Everything else circles in an orbit, but Sagittarius A* stays still.

Astronomers have made careful measurements using radio and infrared telescopes to estimate both the mass and size of Sagittarius A*. The results are amazing: Sagittarius A* is millions of times heavier than the Sun, but is so small that it would fit inside the orbit of Mercury. No normal star can be this massive but this small. Instead, astronomers now believe that Sagittarius A* is almost certainly a supermassive black hole—a giant, condensed, collapsed cloud of material whose density is so high and whose gravity so intense that all matter that strays too close is sucked into its maw. As the name "black hole" suggests, not even light can escape from a black hole's interior. But while the Milky Way's central black hole is itself presumably black, gas on frantic and perhaps final orbits, very close to its rim, produces large amounts of light and heat, producing the source we see as Sagittarius A*.

But what does a supermassive black hole at the center of the Milky Way, the one object in the entire Galaxy that is perhaps not moving at all, have to do with hypervelocity stars?

The answer first began to emerge in 1988, when American astronomer Jack Hills considered what might happen when two tightly bound stars, orbiting each other in a binary system, have a close encounter with a supermassive black hole. Sometimes both stars will have a near miss with the black hole and will continue on their way. At other times, both stars will fall into the black hole and disappear forever. But what Hills realized was that, occasionally, the black hole's gravity will be able to break the binary system into two separate stars. One star will then be captured by the black hole, but the other can be catapulted away at extremely high speeds of well beyond 600,000 miles per hour.

As spectacular as such events would be, unfortunately it's very difficult to find any stars that have been ejected from the galactic core in this way. First, these hypervelocity stars are expected to be quite rare—one hypervelocity star might be generated in this way only every 100,000 years. Furthermore, these stars will not only be moving fast enough to avoid falling into the central supermassive black hole, but they will have sufficient speed to escape the gravitational clutches of the entire Milky Way. Less than 100 million years after being slingshot out of the galactic center (a cosmic blink of an eye), a hypervelocity star will leave the Milky Way altogether, and will begin traversing the empty reaches between galaxies.

The only hypervelocity stars we have a chance of finding are the ones that were expelled by Sagittarius A* relatively recently (the others have all long since escaped the Milky Way). Since such stars are shot out comparatively infrequently, we expect there to only be a small number that we might be able to see, perhaps a thousand such stars spread over the entire sky. Finding this handful of stars among the hundreds of billions of other ordinary stars in the Milky Way is the ultimate needle-in-the-haystack problem. After Jack Hills published his calculation, most astronomers regarded the concept of hypervelocity stars as an interesting and reasonable idea, but one that might be too hard to ever put into practice.

That all changed in 2005, when a young American astronomer, Warren Brown, set out to make a new map of the mass and structure of the Milky Way. As part of this project, Brown was using a telescope in Arizona to accumulate a vast catalog of previously unstudied stars. Most of the stars seemed normal, but one faint star known as "SDSS J090745.0+024507" stood out.

Through the Doppler effect that I mentioned earlier, motion toward or away from us can not only shift the pitch of the siren on a passing ambulance, but can also change the color of a fast-moving star. A star moving toward us will have its color shifted to the blue end of the spectrum, while a star moving away will be shifted to the red end. And while it is routine to see a star whose color is shifted slightly bluer or slightly redder due to its particular motion, SDSS J0907 was much, much redder than it should be. To a casual observer, its colors seemed quite normal, but Brown's careful measurement of its Doppler shift showed that SDSS J0907 is moving away from the Sun at more than 1.9 million miles per hour. Some of this motion is due to the Sun's own orbit around the Milky Way: Subtracting this, we find that the motion of SDSS J0907, as viewed from a "stationary" vantage point, would be about 1.5 million miles per hour.

This incredible speed alone suggests that SDSS J0907 was blasted out of the center of the Galaxy by a close encounter with Sagittarius A*—it's hard to come up with any other way of getting an otherwise ordinary star to travel this fast. But the smoking gun is the direction in which SDSS J0907 is moving—its trajectory is aimed directly away from the center of the Galaxy. Not only is this star traveling faster than any other, but the starting point of its journey seems to have been exactly where Jack Hills predicted—a close encounter with the giant black hole at the Milky Way's core.

Calculations suggest that SDSS J0907 and its binary companion approached Sagittarius A* about 150 million years ago, when dinosaurs still dominated the Earth. Its companion star fell into the black hole and was lost forever, but the gravity of the black hole gave SDSS J0907 a colossal boost in speed, and it sped outward at more than a million miles per hour, never to return. So far SDSS J0907 has traveled more than 400,000 light-years from Sagittarius A*, and shows no sign of slowing. Already on the outskirts of the Milky Way, SDSS J0907 easily has enough speed to completely escape our Galaxy, and before too long will enter the empty wastelands of intergalactic space.

Warren Brown's remarkable discovery has created an

entirely new field of study. Over the last few years, Brown and his team have been doggedly searching for more hypervelocity stars, while other astronomers have begun to run sophisticated simulations to flesh out the details of Jack Hills's original idea. The rapidly growing tally currently stands at around 20 high velocity stars, each speeding at more than 600,000 miles per hour away from Sagittarius A*. SDSS J0907 currently still holds the record for the fastest of this group, but it will be no surprise if an even faster star is soon discovered.

As we begin to build up an entire catalog of these remarkable stars, along with how fast they are moving, which direction they are heading, and how long ago they were shot out of the center of the Milky Way, the hope is that we can develop a fossil history of hypervelocity stars, revealing the aggressive treatment that our Galaxy's central black hole has doled out over the last several million years to those who have strayed too close.

Just for kicks

Incredibly, hypervelocity stars are not the fastest stars known. That record goes to neutron stars, the tiny, rapidly spinning stars that we met in chapters 1 and 3, left behind as collapsed cores when massive stars end their lives in supernova explosions.

In these catastrophic events, a star blasts off its outer layers into space, releasing enough energy in a fraction of a second to temporarily outshine an entire galaxy. If a supernova explosion were perfectly spherical and symmetrical, debris would move out evenly in all directions, and the newly produced neutron star would sit stationary at the center. As an analogy, consider a game of pool, which begins with 15 numbered balls arranged in a pyramid, and the black eight ball at the center. If the first player breaks in just the right way, 14 of the 15 balls in the rack will fly off in all directions, but the black ball will remain untouched.

However, supernova explosions are not good pool players. For reasons that we are still struggling to understand, these detonations are not symmetric, but rather material is blasted outward in some directions faster than others. Even if the asymmetries involved are microscopic, the energy of the explosion is so large that this can kick the newborn neutron star in a random direction at extreme speed. Indeed, the typical speed with which young neutron stars are ejected from supernova explosions is well above 700,000 miles per hour. Even an ordinary, unremarkable neutron star can comfortably match or even outpace the most extreme hypervelocity stars.

Measuring accurate speeds for neutron stars is tricky. They are extremely dim objects without well-defined colors, so the Doppler-shift technique used for hypervelocity stars does not work. Instead, to measure the velocities of neutron stars we need to be far more direct: We simply measure their positions in the sky, wait a few months or years, and measure their positions again. If the position has changed, and if we know the distance to the star, then we can work out how fast the star is moving.

While this sounds simple, this technique, known as a "proper motion measurement", can be very challenging: Even for motion at very high speeds, stars are at such great distances that they barely seem to move at all as viewed from Earth. For example, a neutron star 10,000 light-years away, moving at 700,000 miles per hour, would take 100,000 years to move across the sky by a distance equal to the diameter of the full Moon. Exquisitely precise measurements are required to see such motion in a time span short enough to be useful.

Fortunately, modern telescopes are up to the task: Such determinations, while not simple or straightforward, are very much feasible. As a consequence, proper motions have been determined for more than 200 neutron stars, and some of their speeds are staggering. At the moment, the fastest known neutron star, and indeed the fastest known star of any kind, is a star in the constellation Puppis known as "RX J0822.0-4300," located about 7,000 light-years away. As viewed from Earth, the supernova that gave birth to RX J0822 occurred only 3,700 years ago, and indeed the expanding cloud of gas and dust from the explosion can clearly be seen, surrounding the star on all sides.

RX J0822 is not an especially fast-spinning neutron star. While in chapter 3 we were confronted by a neutron star spinning 716 times per second, RX J0822 rotates at the comparatively sedate rate of nine times per second. However, what RX J0822 lacks in spin it makes up for in sheer speed. Astronomers made repeat measurements of the position of RX J0822 in 1999, 2001, and 2005, using the ultra-sharp vision of NASA's Chandra X-ray Observatory. The shift in the star's position between 1999 and 2005 was minute: equivalent to the width of a human hair viewed from 60 feet away. As puny as this sounds, this is an impressive effort for a star thousands of light-years away. The velocity of RX J0822 as measured by our cosmic radar gun? An incredible 3.5 million miles per hour.

It is hard to imagine just how fast this is: 4,700 times the speed of sound, 150 times faster than the *Apollo 10* astronauts, 50 times faster than the orbit of the Earth around the Sun, 7 times faster than the orbit of the hot Jupiter WASP-12b, and more than twice as fast as the hypervelocity star SDSS J0907. Every 2.5 seconds RX J0822 travels the distance from Los Angeles to New York, and every 4 minutes the distance from the Earth to the Moon.

Remarkably, the supernova explosion that gave RX J0822 its tremendous kick needed to have been asymmetric only at the 3 or 4% level to produce this extraordinary velocity. The extreme speed of a neutron star is only an afterthought to the far more dramatic and energetic cataclysm that is a supernova.

Measuring the high speed of RX J0822 required patience: It took 6 years for this star to shift position by enough that we could measure it. However, neutron stars move so fast that there is another, less accurate, but far more immediate and dramatic way of measuring their motions.

The gas between the stars is rarefied and tenuous. But as a neutron star hurtles through this material at high speed, it piles up gas in front of it like snow in a snowplow. This gas is heated to a temperature of thousands of degrees, and glows brightly. As the neutron star sweeps past, some of this gas then trails behind in a fiery wake. The result is called a "bow shock," and around a dozen such objects have now been identified.

While bow shocks give us only reasonably rough estimates of neutron star speeds, they provide further evidence that these tiny objects move extremely quickly. And like distant kites swirling in the breeze, they allow us to study interstellar gas that would otherwise be invisible.

Just missing light speed

A speed limit posted by the side of the road is a recommendation and a guideline, but it is not a hard, unbreakable rule. You can go faster without difficulty if you choose (although a speeding ticket may quickly follow).

While there are no traffic cops or radar guns in space, incredibly there *is* a cosmic speed limit. And in this case it is ruthlessly enforced—no moving object can ever exceed this speed within the laws of known physics. This universal speed limit, of course, is the speed of light: 186,282.397 miles per second, or just over 670 million miles per hour.

Albert Einstein, in his Special Theory of Relativity, predicted that as objects try to accelerate toward the speed of light, they get increasingly heavier, and so require more and more power to keep accelerating. The closer you approach the speed of light, the more energy you need to go any faster. Try to push right up to light speed, and the amount of fuel and energy required ramps up to infinity. No matter how hard you try, you can never quite get up to light speed. As bizarre as this seems, careful experiments using particle accelerators have confirmed that moving objects really do experience these effects, and that indeed nothing can travel faster than the speed of light.

This is of little consequence for all the remarkably fast-moving planets and stars that we've discussed above—none of them get anywhere near the speed of light. Even the 3.5 million miles per hour that astronomers have clocked for the neutron star RX J0822 is a mere 0.5% of light speed. But the Universe is filled with mysterious particles called "cosmic rays," which move much, much more quickly than RX J0822, and which can push the light barrier right to its limits.

The French scientist Henri Becquerel discovered radioactivity in 1896. Soon after, scientists realized that weak, background radioactivity was all around us. It was thought that this radioactivity was emanating up from the ground, from uranium and other naturally occurring radioactive elements. However, in 1912, the Austrian physicist Victor Hess performed a clever experiment in which he carried a radioactivity detector up to high altitudes in a hot-air balloon. As he ascended, the level of radioactivity at first began to drop, because the Earth's atmosphere was absorbing radioactive signals coming up from the ground. However, as the balloon continued to rise, the level of radioactivity began to go up again. By the time Hess was three miles up, the level of radioactivity was four times higher than he had been seeing at ground level. Hess concluded firmly that at least some of the naturally occurring radioactivity that we experience every day was coming not from rocks and minerals buried beneath us, but from space. Hess received the 1936 Nobel Prize for the discovery of these mysterious cosmic rays.

One hundred years after Hess's pioneering work, we now know quite a lot about these cosmic rays. We know that they are not "rays" at all, but mostly protons and other subatomic particles. We know that the entire Milky Way is flooded with them. And we know that they travel very, very fast.

Trillions of cosmic rays crash into the Earth every second. Most are produced by energetic outbursts on the surface of the Sun, and typically travel at around 99% of the speed of light. This is faster than almost anything else in the Universe, but is still almost 7 million miles per hour slower than light itself.

However, there is a tiny, tiny fraction of cosmic rays that makes 99% of light speed seem positively sluggish. This rare population, known as "ultra-high-energy cosmic rays," goes all the way up to the fastest speeds possible under the laws of physics.

The definitive record for the fastest speed ever measured in the Universe, except for light itself, was set at precisely 1:34:16 a.m. on Tuesday, October 15, 1991, near the small town of Dugway, Utah. At this moment, a cosmic ray, probably a proton, slammed into the Earth's atmosphere, detonating into a spectacular shower of sparks. A telescope named "Fly's Eye," which had been specially designed for this purpose, caught a fleeting glimpse of this shower. Using the pattern and extent of these sparks, the scientists operating Fly's Eye were able to reconstruct the speed at which this proton must have hit us, and the result was astonishing. Before it was smashed to pieces in the atmosphere, it was moving at 99.99999999999999999996% of the speed of light! Or to put it another way, suppose that this proton and a light ray had a race over a length of 1,000,000 light-years. The light ray would win, but only just. After 1,000,000 years neck and neck, the light ray would beat the proton to the finish line by about 1.5 inches. Talk about a photo finish.

While most cosmic rays seen by Fly's Eye and other telescopes are referred to only by their catalog numbers, the one seen in October 1991 has its own name: For understandable reasons, cosmic-ray scientists refer to it as the "Oh-My-God Particle."

At some point millions of years ago, in some distant galaxy, the proton that would one day become the Oh-My-God Particle was probably an ordinary, unremarkable atomic particle, drifting about space like any other. However, something then happened to this proton to boost it up to its unimaginable speed. We don't know how this happened, but the energy involved is staggering. When the Oh-My-God Particle arrived at Earth, it had more than 12 calories of energy.

This might not sound like much: You burn more calories than this walking to the bus stop. But let's put it in perspective. The Large Hadron Collider is the most powerful particle accelerator ever constructed—building it has cost more than 9 billion dollars, and running it requires several hundred megawatts of electricity. But the Large Hadron Collider can boost subatomic particles only up to a maximum energy of around 200 billionths of a calorie. Somehow, some naturally occurring process in the cosmos can take a single proton, and can give it 50 million times more energy than we humans are capable of.

It's important to realize that the Oh-My-God Particle can't be passed off as a one-off, freak event. While we've never seen another cosmic ray at these speeds, this is only because we probably haven't been looking hard enough—current estimates suggest that a cosmic ray moving as fast as, or even faster than, the Oh-My-God Particle collides with the Earth's atmosphere somewhere on the globe about once every 20 seconds.

Nothing motivates astronomers like a cosmic mystery that needs to be solved. And so there is now a large international effort, based around a giant cosmic-ray experiment known as the Pierre Auger Observatory, whose main aim is to find other speed demons like the Oh-My-God Particle. Auger has two sites: a Southern Hemisphere one in Argentina, and a Northern Hemisphere one to soon begin construction in Colorado.

Auger consists of specially designed telescopes, spread over more than a thousand square miles. These telescopes stare at the sky every night, looking for the telltale flash of light that marks the air shower produced by an incoming cosmic ray. So far the record set by the Oh-My-God Particle has been challenged but not yet broken—the fastest particle seen by Auger, detected in Argentina on January 13, 2007, was moving at 99.99999999999999998% of the speed of light and had an energy of 5 calories. But Auger is playing a waiting game: It has only been taking data since 2004, and it's just a matter of more time, and more data, before something even more spectacular than the Oh-My-God Particle reveals itself. Auger's ultimate goal is to detect not just one, but many of these incredible cosmic bullets. If we find enough of them, and see which direction they have come from, then just like a ballistic expert, we can hopefully figure out what is firing them.

6

EXTREMES
OF MASS

"Mass" and "weight" are everyday ideas. These two words are so familiar and easily understood that we use them as shorthand to convey a wide set of ideas and emotions. For example, a sad event can leave us with a "heavy" heart. A challenging task might be seen as a "massive" undertaking.

But despite the familiarity of mass and weight, it took scientists hundreds of years to begin to develop an understanding of these concepts. Even now, there are subtle aspects to these ideas that we continue to puzzle over.

Imagine a child and an adult, each sitting cross-legged on a backyard trampoline. The trampoline sags down toward the ground due to their presence: The child makes a small dent in the trampoline's surface, while the adult makes a much larger one.

Now let's suppose we move the trampoline to the surface of the Moon (and let's make sure the trampoline is inside a giant air dome, so that our trampolinists don't require space suits or oxygen). Gravity on the Moon is about one-sixth that of Earth, so if the adult and the child sit on the trampoline now, they will not sink down anywhere near as much. However, the depression made by the adult will still be larger than the one made by the child.

This simple picture simultaneously shows us both mass and weight. The adult has more mass than the child, so will always make a bigger dent in the trampoline's surface. But both the adult and child make bigger dents on Earth than they do on the Moon—a person's weight depends on the pull that gravity exerts on them, even though their mass is unchanged. The extreme situation is an astronaut in orbit, who floats around with no weight at all, but whose mass is the same as on Earth (much more about this in chapter 9).

So weight is something that depends on where you are: On the Sun or Jupiter, you would weigh a lot more than you would on Earth, but on Mars or Mercury, you would weigh considerably less. On the other hand, mass is an intrinsic property of an object. Wherever you are in the Universe, your mass at any given moment is always the same. Mass is something innate, related to the atoms and molecules of which you're constructed.

With this understanding in hand, we're ready to face the enormous masses found throughout the cosmos. A word of caution though: Even though a physicist will carefully define weight and mass as two distinctly different concepts, the English language isn't so accommodating. In what follows, I'll only be considering the masses of different bodies throughout the Universe, not their weight. Nevertheless, as all of us do every day, I'll still talk about "weighing" objects, since "massing" them doesn't make a whole lot of sense.

Once we understand the difference between mass and

weight, extremes of mass are something that we are reasonably good at understanding. An elephant weighs several million times more than an ant, but we have no trouble visualizing both these masses, plus the whole range in between. Even among humans, there is a huge range of mass: A healthy newborn baby weighs just a few pounds, while the heaviest sumo wrestlers can tip the scales at more than a quarter of a ton.

However, once we move beyond our everyday surroundings and into the cosmos, the situation quickly becomes hard to comprehend. Even the smallest known asteroid, tiny 2008 TS26 that we met in chapter 4, weighs hundreds of pounds. In comparison, the largest asteroid in our solar system, Vesta, is 300 miles across, and weighs about 290,000 trillion tons.

Let's keep going. Mercury, the smallest planet in the solar system, weighs 360 million trillion tons. Our own planet, Earth, weighs 6.5 billion trillion tons. Jupiter weighs more than 2 trillion trillion tons, which is more than double the mass of every other planet in the solar system combined. And the Sun, whose gravity holds the whole solar system together, is a thousand times heavier again, with a staggering mass of 2,200 trillion trillion tons.

Enormous numbers like these are impossible to grasp in any meaningful way, and yet we haven't even yet left the solar neighborhood. If we step into the beyond, we quickly find that the Universe abounds with swarms of unexpectedly lightweight stars, along with far, far heavier creatures than even our gargantuan Sun.

Little but long-lived

The Sun is unimaginably massive, but is it a typical star? Or is it noticeably heavier or lighter than the average?

Before we can answer this, we need to ask an even more fundamental question: How do you possibly determine the mass of a star? In the everyday world, we can weigh something simply by putting it on some scales. But stars are extremely large and very distant, so this approach is not very practical.

The great English scientist Sir Isaac Newton gave the answer to us in 1687, through his universal law of gravitation. Newton wrote that any object with mass exerts gravity, and that any object with mass experiences gravity. What's more, the strength of the gravitational force between two objects depends on the mass of each object and on the distance between them. This simple but remarkable statement has many implications.

For example, not only does the Earth exert a gravitational force on you, pulling you downward, but you exert an equal gravitational force on the Earth, pulling it upward. Although the two gravitational forces are equal, the Earth's pull on you has a much larger impact than your pull on the Earth, because the Earth's mass is very large while yours is comparatively small.

Another example: The Earth's pull isn't the only gravity each of us experiences, but rather we continuously experience the gravitational pull of everything around us. This book that you're reading, the chair you're sitting on, the plane flying overhead, and the person walking by across the road are all exerting gravity on you. You don't feel these effects though because the masses of these objects are so small—the Earth's gravity overwhelms these other small influences.

And finally, every other object in the Universe exerts its gravity on you too. All the objects we've discussed in this book—distant dark nebulas, enormous elliptical galaxies, and fast-spinning neutron stars—are pulling on you with their gravity. In these cases, the masses involved can be very large, but the distances to them are huge. Again, the Earth's gravity wins out.

So here on Earth, we don't normally see the full, glorious effects of Newton's law of gravity in action. Gravity means that apples fall from trees, people bounce up and down on trampolines, and we all ultimately stay pretty firmly fixed to the ground. But beyond the bounds of Earth, gravity truly shows its elegant, universal appeal.

Many stars come in pairs, known as "binary systems." The two stars in a binary are bound to each other by Newton's law of gravitational attraction, and swirl around each other in circular or oval-shaped orbits for millions or billions of years. What's more, Newton's equations tell us exactly how the behavior of the two stars' orbits depends on their masses. Specifically, if we can measure both the average distance between two stars in a binary and the length of time it takes them to orbit each other, we can then accurately determine both their masses. For example, if two stars are widely separated but orbit rapidly, then this means they must both be especially massive. On the other hand, if the stars in a binary are close together but orbit slowly, then they must both be light.

If we use this technique, weighing stars becomes relatively straightforward. We just need to find a pair of binary stars and watch them in their orbits. Provided we also know the distance to the system (which can sometimes be simple to determine or sometimes quite challenging, but that's a whole other story . . .), we then have the duration of the orbit, the separation of the two stars, and hence their masses.

Unfortunately, there are complications. If the stars are far enough apart that we can easily measure their separation, then this probably also means that their orbit is very slow. A binary with a million-year orbit is not a practical system to study! On the other hand, if the orbit is fast enough for us to track it in a reasonable time, then the stars might be too close together for us to measure their separation, or to even realize that there are two stars in the system rather than just one.

Astronomers thus use a whole range of tricks to study binary stars and to calculate their masses. Sometimes the stars are too close together to separate even with a powerful telescope, but we can use the Doppler effect discussed in chapter 5 to see the tiny shifts in each star's color to the blue or to the red as they each repeatedly swing toward us or away from us in their orbit. This telltale change in color can give us much of the information we need to calculate the orbital properties and therefore the two stars' masses.

At other times, the orbit will be almost exactly edge-on, so that one star passes in front of the other and causes an eclipse once every orbit, similar to the transit method used for finding exoplanets that I explained in chapter 5. Or in cases where one star is bright but the other is too faint to see, we can simply see the bright star wobble back and forth as it orbits its unseen companion.

But whichever approach one adopts, the common theme is

that binary stars are the lynchpin of all measurements of stellar masses. For a star that's not in a binary system, there is often no way to measure its mass. Instead, we note the particulars of the star's color, temperature, and brightness, and try to compare it to another star with similar properties, but in a binary system and with a known mass.

So now let me ask my initial question again. Is the Sun excessively heavy compared to the average star, embarrassingly puny, or reasonably average?

Imagine you are in high school, taking an exam. You are surrounded by hundreds of other students, all furiously writing out their answers before time expires. You gaze up from your notes and look around you, and it suddenly strikes you that every single student in the room is left-handed.

This would be a very unlikely and unexpected situation, given that we all know that the vast majority of people are right-handed. But this is exactly the strange scenario we face every time we stargaze.

If you live in a big city, where the night sky is never especially dark, then you might only be able to see a few hundred stars with your naked eye. Almost every one of these stars has a mass larger than the Sun's, and you thus might be tempted to conclude that the Sun is one of the smallest stars in the Milky Way. However, like the roomful of left-handers, what you are seeing is a biased and unrepresentative view of the typical population. The stars you can easily see are all quite heavy, but they are also very good at attracting attention as a result of their extreme brightness.

The true situation, somewhat surprisingly, is that the Sun is the galactic equivalent of a sumo wrestler, with a mass far in excess of what is typical for most of the population. But to begin to see even the tip of the iceberg of "normal" stars, you have to look quite hard.

If you have good eyes, then on a clear, moonless night in the constellation of Cygnus you might just be able to make out the faint orange star known as "61 Cygni." 61 Cygni is actually two similar stars in a binary, in an orbit taking almost 700 years, and with an average separation of 8 billion miles. Using this information, astronomers can calculate the masses of both stars: The heavier of the pair, known as "61 Cygni A," weighs in at 70% of the Sun, while its slightly smaller companion, "61 Cygni B," has 63% of the Sun's mass.

61 Cygni is notable because these are the two least massive stars visible to the naked eye. But even the stars in 61 Cygni are somewhat heavyset compared to the norm. The overwhelming majority of stars, about 85% of all the stars in the Milky Way, are the long-lived red dwarfs that we discussed in chapter 3. These stars, weighing just 10–40% of the mass of the Sun, lurk in every corner of the Galaxy, but usually escape notice because of their faint and feeble appearance.

As dim and unimpressive as red dwarfs might be, they are very much full members of the stellar fraternity. Just as for the Sun, nuclear fusion reactions take place in a red dwarf's interior, steadily converting hydrogen into helium, and giving off heat and light as a by-product. A key difference in this process is the temperature in the star's interior. At the center of the Sun, the temperature reaches around 27,000,000°F (see chapter 1). But for a red dwarf, the smaller mass means a reduced pressure in the core, so that the central temperature for a red dwarf is less than half of that for the Sun. The rate at which fusion proceeds in a star's interior is extremely sensitive to the temperature, meaning that red dwarfs burn their fuel very slowly, making them cool, red, and dim. Even the most energetic red dwarfs generate about one-tenth the energy of the Sun.

As you would expect, the lightest red dwarfs are the feeblest, their internal nuclear reactions taking place at a snail's pace because of their low core temperatures. However, this situation cannot extend downward in mass indefinitely. For core temperatures below about 9,000,000°F, nuclear fusion cannot proceed. The very definition of a normal star requires nuclear reactions to be taking place in its interior, so there is a firm lower limit on the coldest, and hence lightest, an object can be and still be called a star. Objects of lower mass certainly exist: These cool balls of gas are known as "brown dwarfs." But brown dwarfs generally do not shine or burn fuel in the way that stars do.

Given this situation, we can then ask the question: What is the lightest possible mass a star can have? Astronomers have invested considerable effort to calculate this number. The exact value is still a matter of debate, and also depends on the specifics of the star's composition (for example, almost pure hydrogen, or containing significant impurities from hydrogen and other gases). However, the broad consensus is that the lightest a star can possibly be is about 150 trillion trillion tons. This is about 23,000 times heavier than the planet Earth, and so by our standards is still stupendously large. However, it is a mere 7% of the Sun's mass—by any astronomical measure, this is extremely lightweight indeed.

With this 7% threshold in mind, we can then hunt through the night sky to see if there are indeed any stars of this low mass. The search is a challenging one, not only because such stars will be faint and well hidden, but because when viewed from a distance of many light-years, it becomes very hard to determine whether such an object is a red dwarf just above the mass threshold to be a star, or is a brown dwarf just below the cutoff.

It is therefore difficult to be definitive as to what is the lightest known star, since any such claim might be countered with the argument that this object is instead the heaviest known brown dwarf. With this caveat in mind, one likely candidate for the lightest star yet discovered is an object known as "GJ 1245C," which sits at a distance of 14.8 light-years in the constellation of Cygnus. GJ 1245C was discovered in 1984, and is part of a binary system with an orbit of 15 years and an average separation of 330 million miles. The most recent careful measurements of its mass, carried out by American astronomer Todd Henry using the Hubble Space Telescope, show that GJ 1245C has a mass only 7.4% that of the Sun, just above the expected minimum limit for a star. We are yet to obtain absolute confirmation that GJ 1245C is a red dwarf, but it certainly fits the bill from the information we currently have at hand.

GJ 1245C is especially unimpressive. It is about 10,000 times too faint to see with the naked eye, and is barely visible even through a modest-sized telescope. If we were to put GJ 1245C at the center of our solar system it would be barely brighter than the full Moon, while as a result of its weak illumination, the Earth's average surface temperature would drop to a chilly −400°F!

Objects like GJ 1245C might be puny and pathetic, but they will certainly have the last laugh. These scrawny, dim red dwarfs outnumber ponderous stars like the Sun by a factor of 10 to 1. And GJ 1245C is so miserly at burning its fuel that it will shine for a trillion years, outliving the Sun by a huge factor. The meek will not only inherit the Earth, but the rest of the cosmos as well.

As big as they come

Although most of the stars in the sky are small and unimpressive red dwarfs, we have already seen in chapter 4 that it's the rare, large stars that catch our attention. As I mentioned above, almost every bright naked-eye star in the sky is heavier than the Sun. For example, the brightest star in the sky, Sirius, has a mass almost exactly twice that of the Sun, while the second-brightest star, Canopus, has more than eight times the Sun's mass. But even these big beasts are unremarkable by galactic standards.

Let's start with the heaviest star we can see unaided. That title most likely belongs to Alnilam, the middle star in Orion's Belt. Alnilam is not a binary star, but by comparing it with other stars with accurate mass measurements we can estimate that it weighs an astonishing 90,000 trillion trillion tons, some 40 times more than the Sun. Alnilam is also hundreds of thousands of times more luminous than the Sun, but is 1,300 light-years away. This is much more distant than almost all other naked-eye stars (Sirius is 8.6 light-years away; Canopus is about 300 light-years distant), so that instead of being overwhelmed by Alnilam's brightness, we see it merely as the 30th brightest star in the sky.

This extreme level of stellar obesity is unusual, but it still does not come close to pushing the limit. Since more massive stars are increasingly hotter and brighter, one might expect that the heaviest stars in the Galaxy should be devastating in their brilliance, and easy to spot. However, finding the super heavyweights of the stellar community is extremely challenging. Not only do we expect these stars to be very rare, but we anticipate that they will also normally be especially well hidden.

The heaviest stars gobble their fuel extremely quickly, and so are extremely short-lived, lasting just a few million years from birth to death. This means that they have no time to drift around the Galaxy during their lives, and will inevitably be found very close to their places of birth. These stellar nurseries are complicated, messy places, filled with many bright young stars, glowing gas, and dark nebulas (see chapter 2). If we are hunting for the heaviest stars, such agglomerations are the obvious places to look. However, with all this other clutter obscuring our view, finding the most massive stars is not simple. First, we have to pick which of the hundreds of stars in the region we think might be the heaviest, and then we have to accurately measure its mass despite all the other stars and gas in the way. Finding ultra-heavy stars is thus a painstaking process, requiring special effort.

To narrow down the search for the most massive star in the Milky Way, I need to introduce you to an extremely rare class of objects known as "Wolf-Rayet stars." Named after the French astronomers Charles Wolf and Georges Rayet (who discovered these remarkable stars in 1867), Wolf-Rayet stars are very hot, very luminous, and very heavy. Their extreme temperatures cause them to gradually evaporate, so that their outer layers stream out into space at a furious rate. For a typical Wolf-Rayet star, this "stellar wind" travels out from the star at speeds well in excess of 6 million miles per hour, and causes the star to lose weight at a rate of 650 trillion tons every second.

All this fury and energy is short-lived, however. A Wolf-Rayet star's high temperature causes it to burn through its fuel extremely quickly, while at the same time the star sheds a substantial fraction of its mass through its strong wind. After less than a million years (a blink of an eye compared to the 10-billion-year life span of the Sun), a Wolf-Rayet star exhausts its fuel supply, collapses under its own gravity, and then dies in a catastrophic supernova explosion. Because of both their unusual nature and their short lifetimes, Wolf-Rayet stars are a rare breed. Only about 300 Wolf-Rayet stars have been identified in the Milky Way so far; the total number still to be discovered is thought to be around 6,000–8,000.

After this introduction, it will come as no surprise that almost all the contenders for the Galaxy's heaviest star are Wolf-Rayet stars. To track down the current record holder, we need to turn our attention to NGC 3603, a beautiful glowing mixture of gas, dust, and newborn stars in the constellation of Carina. Deep within NGC 3603, buried amid the light from dozens of other bright stars, is an extraordinary system known rather simply as "A1." Through painstaking measurements, astronomers have discovered that A1 is actually two Wolf-Rayet stars, 25 million miles apart and orbiting each other every 90.5 hours. Because A1 is a binary, we can apply Newton's laws of motion just as for tiny GJ 1245C described earlier. Astronomer Olivier Schnurr and his colleagues did just that in 2008, and the result is extraordinary. The smaller star of the pair outweighs the Sun by a factor of 89, and by any measure would be an incredible object. However, it is completely overshadowed by the larger star of A1, which is a spectacular 116 times heavier than the Sun, with a mass of around 250,000 trillion trillion (250,000,000,000,000,000,000,000,000,000,000) tons!

At the time of this writing, this behemoth was the heaviest known star in the Milky Way with an accurate and reliable mass measurement, but that record is not likely to stand for long. There are strong suspicions that other stars, yet to be put on the scales, are even heavier. It's also important to bear in mind that because of the rapid mass loss that Wolf-Rayet stars experience, the birth weight of these stars is potentially much higher than the mass we might measure today. An example is the star known as "WR 102ka," located about 26,000 light-years away toward the constellation Sagittarius. WR 102ka is another enormous Wolf-Rayet star, but it is so deeply buried in the dust and gas of its surroundings that as yet we can only crudely estimate its mass, which seems to be about 100 times that of the Sun. However, based on its current temperature and brightness, astronomers have calculated that WR 102ka probably was 150 times heavier than the Sun when it was born. If it were not for its self-enforced drastic weight-loss program, WR 102ka perhaps would reign supreme as the Milky Way's heaviest star.

Is there any limit to how heavy a star could ever get? In our Galaxy, it is probably difficult to produce a star weighing more than 150–200 times the Sun's mass. For anything heavier, such a star's strong wind and other related effects would probably tear it apart before it could fully form and begin shining.

However, much earlier in the Universe's history, the situation perhaps was quite different. The process by which a star forms and evolves depends in a complicated way on the environment in which it was born. In particular, as we discussed in chapter 3, the stars forming today have a substantial metallicity. That is, they have significant levels of elements heavier than hydrogen or helium. These impurities regulate the processes through which a star collapses from a cloud of cold gas, burns its fuel, and potentially loses its mass through a stellar wind.

In chapter 3 we considered the as-yet-undiscovered category of Population III stars: the very first stars in the Universe to form, many billions of years ago. Population III stars are expected to have had very different properties from the stars we see today, because of their essentially zero metallicity. In particular, astronomers have calculated that Population III stars could have been extremely massive indeed, extending way past any Milky Way star, perhaps up to 300–500 solar masses or even beyond. While any such titanic beasts would long since have burnt out their fuel, upcoming new telescopes will be able to look out into space and back into time to hunt for these distant, massive objects. One thing is for sure: Any heavyweight championship record set long ago by one of these Population III stars will almost certainly stand the test of time.

The center of the action

We have spanned the vast diversity that the stars in the Milky Way can offer, from the featherweight red dwarf GJ 1245C to the supermassive Wolf-Rayet star A1. But there are other things in our Galaxy besides stars, and many of these are far heavier. For example, the Milky Way is studded with the dark nebulas that we discussed in chapter 2. These huge clouds of dust and gas are far more massive than any single star, the largest ones weighing more than 100,000 times the mass of the Sun.

But the debate over the heaviest single object in the Milky Way is an easy one to settle. Without any question, that title goes to Sagittarius A*, the giant black hole at our Galaxy's core, which we met in chapter 5.

The technique astronomers use to weigh stars, through their orbit around their binary companion, won't work for Sagittarius A*, because it is at the absolute center of the Milky Way, and is not in an orbit around anything. But since almost everything else in the Milky Way orbits around Sagittarius A*, we can apply related techniques to the binary-star method in order to estimate its mass. In particular, astronomers have been able to identify dozens of stars deep in the Milky Way's interior, which circle Sagittarius A* in very tight, fast orbits. The closest known star to Sagittarius A* takes just 16 years to complete a circuit, compared to the more than 200 million years that the Sun takes to do so. These high-speed thrill seekers are the likely forebears of the hypervelocity stars that we discussed in chapter 5.

Precise observations of the shapes of the orbits for these fast-moving stars have allowed astronomers to make surprisingly accurate measurements of the mass of Sagittarius A*. The result? Sagittarius A* far outweighs any mere star, clocking in at 4.3 million times the mass of the Sun, or more than 9 billion trillion trillion tons.

But on a cosmic scale, even Sagittarius A* is a comparatively puny object. Studies of the night sky have revealed that most other large galaxies also seem to harbor supermassive black holes at their centers. In most cases we cannot weigh these other black holes anywhere near as precisely as we can for Sagittarius A*, because our telescopes do not have the sharpness of vision to see individual stellar orbits around the central black hole as is the case for the Milky Way. Instead, we need to resort to more indirect methods. For example, the level of activity and radiation in the immediate surroundings of a supermassive black hole seems to be crudely linked to its mass. So if we can measure how intense the light is close to a central black hole in a galaxy, we can estimate how heavy that black hole is likely to be.

While we give up our ability to make accurate measurements in this way, we now open up tens of thousands of other galaxies to study. And when we do so, we find that Sagittarius A* is an absolute pipsqueak. Supermassive black holes more than 100 times heavier than Sagittarius A* are routine: There are a myriad of galaxies in the sky whose central black holes have masses that easily exceed a billion times the mass of the Sun.

Because of the uncertainty in the mass measurements of these distant objects, it is difficult to be definitive as to which one is the heaviest. However, a genuine candidate for the title of heaviest black hole in the Universe is an object known as "S5 0014+813," which is 12 billion light-years away in the constellation of Cepheus. Its mass, as calculated by a team led by Italian astronomer Gabriele Ghisellini, is 40 billion times that of the Sun! This extraordinary measurement still awaits confirmation and refinement: As Ghisellini himself has acknowledged, S5 0014+813 is a complicated object and his calculation is far from definitive. Nevertheless, with masses comfortably exceeding 10 trillion trillion trillion tons for the biggest specimens, there is no question that supermassive black holes are the heaviest single objects in the Universe.

Of course, astronomers are not satisfied with confirming the mere existence of these remarkable objects. We also want to know why they are so extraordinarily massive. The defining characteristic of a black hole is its gravity, so powerful that once matter falls inside it, it can never escape. Thus while Wolf-Rayet stars begin heavy but are doomed forever to lose weight as they age, black holes are the complete opposite. A supermassive black hole can only ever gain weight, growing ever more corpulent as it sucks stars and gas down its throat in an insatiable fashion.

This simple fact allows us to make some firm predictions. If we could somehow look into the future, supermassive black holes should be even more massive than at present. And if we could look back into the past, these black holes would all generally be lighter than they are now. While we have no way of gazing into the future, it is no problem for astronomers to peer back into the past, simply by looking at very distant objects, whose light has taken billions of years to reach us.

But when we point our telescopes to the most distant galaxies, and measure the masses of supermassive black holes at much earlier times in the cosmos's history, we don't see what we expect. The most distant and hence the earliest black holes we can currently identify correspond to a time when the Universe was less than 10% of its current age. The supermassive black holes in our neighborhood have had an additional 12 billion years to gorge themselves on their surroundings compared to these ancient specimens, so there should be a very clear difference in mass. And yet we consistently find that supermassive black holes all these billions of years ago have very similar masses to the black holes we see today.

Astronomers continue to debate exactly what this means, but the implication seems to be that supermassive black holes have been only lightly snacking for most of the Universe's history, so that they have not grown appreciably in mass over this time. Conversely, these results also imply that there must have been frenzied feasting at very early times in the Universe's history, at epochs that our telescopes are not yet powerful enough to study. In the near future, we hope to be able to push back even further in time, to a frenetic stage in the cosmos's evolution when black holes were growing rapidly as they sucked in everything around them. To reach weights of billions of Suns in such a short time, it seems likely that black holes swallowed not just stars and gas, but even ate each other. This frenzied cosmic cannibalism is now largely a thing of the past, but its record is forever imprinted upon the Universe through the gargantuan masses of the black holes sitting at the heart of almost every galaxy in the sky.

A gaggle of galaxies

Supermassive black holes are staggeringly heavy. But even the biggest black holes are just a small component of the galaxies in which they reside. S5 0014+813, perhaps the biggest of all black holes, might weigh 40 billion times more than the Sun, but our own Milky Way is even heavier, with a mass of more than a trillion solar masses. Less than 10% of this mass is made up of ordinary material such as stars, gas, and planets, with everything else comprised of a mysterious substance called "dark matter" (for which we embarrassingly have no real understanding or explanation!). And as we saw in chapter 4, there are other galaxies far bigger than our own: The enormous galaxy IC 1101 weighs more than 100 Milky Ways combined.

Yet we can keep going bigger. The largest gravitationally bound systems in the Universe are "galaxy clusters," collections of hundreds of galaxies, all in intricate and complicated orbits around each other, spread over millions of light-years.

(Before discussing the stupendous masses of galaxy clusters, it is important to note that in terms of physical size, there are of course even bigger things than clusters. As we saw in chapter 4, the Sloan Great Wall is 1.4 billion light-years in extent, far larger than any cluster. However, the various galaxies and clusters that make up the Sloan Great Wall are not tied together by gravity, but rather are gradually drifting apart as they each move in their own direction. To be able to meaningfully talk about the mass of an object, we require all the different components of that object to be part of a greater whole, close enough to each other to be held together as a group by each other's gravity. Over sizes bigger than a galaxy cluster, the gravitational attraction is too weak and the velocities of individual objects are too high to hold the whole thing together.)

Let's consider the nearest cluster to us, the Virgo Cluster. It's a mere 60 million light-years away and spans more than 15 times the diameter of the full Moon on the sky. The Virgo Cluster contains well over 1,000 different galaxies, plus vast amounts of dark matter and hot intergalactic gas. Making an accurate mass estimate for the Virgo Cluster is challenging. How do we sensibly weigh something that has thousands of different components, embedded in a bath of dark matter that we can't even see? Again we appeal to Newton's laws of motion, from which we can show that even in a complicated, chaotic system like a galaxy cluster, the speed at which the various galaxies orbit and whirl relates to the total mass. If we can get a handle on the typical velocities of galaxies within the cluster, we can start to make an estimate of the mass. Applying this technique to the Virgo Cluster, we find that the total summed mass of all the galaxies, gas, and dark matter is more than that of 1,000 trillion Suns.

But clusters of galaxies can get much larger than this. Just as supermassive black holes can swallow each other to make even bigger black holes, entire clusters of galaxies can collide and merge, resulting in gargantuan agglomerations. The heaviest galaxy cluster known, and thus perhaps the system with the overall title of heaviest object in the Universe, is a cluster known as "Abell 2163," which sits more than 2 billion light-years away in the constellation Ophiuchus. Abell 2163 contains more than 500 galaxies spread over 15 million light-years, all embedded in gas superheated to a temperature of around 270,000,000°F. Its total mass, as measured by a team led by French astronomer Sophie Maurogordato, is about 9,000,000, 000,000,000,000,000,000,000,000,000,000,000,000 tons or 4,000 trillion times that of the Sun!

All of this might leave you feeling a little lightweight. But if it is any consolation, even Abell 2163, for all its stupendously large weight, makes up only one hundred-millionth of the total mass of the observable Universe. On the grandest scale, even the most massive objects in existence are just drops in the cosmic ocean.

7

EXTREMES
OF SOUND

Our lives are never silent.

I am typing these words in a seemingly quiet room, but if I pause a moment to listen, I can hear the ticking of a clock, the rumble of traffic on a nearby road, the hum of an airconditioner next door, and some snatches of distant conversation from passersby.

Throughout a normal day, I will speak to my friends and family, listen to music, and grumble at a plane flying too low over my house: Sound is both a constant presence and a vital component of our lives.

All these sounds are actually minuscule oscillating variations in air pressure, which travel outward from their source at a thousand feet per second ("the speed of sound").

To hear the sounds produced by pressure waves, we rely on microscopic hair-like structures inside our ears. When pressure fluctuations pass into our ear, they tilt these hairs back and forth. This tilting generates electrical signals, which are then carried to our brain and interpreted as sound. This incredible system is so finely tuned that even when the air pressure changes by one part in 10 billion, we can still hear the corresponding noises.

Considering how full of sound the world is, it might be tempting to think that the rest of the Universe could never compete.

After all, if sound is a pressure wave that needs air in which to travel, the vacuum of space must be completely quiet, right?

Indeed, one of the classic moments in cinema is in the film *2001: A Space Odyssey*, when astronaut Dave Bowman returns to his spaceship so that he can shut down the mad computer, HAL. When Dave first enters the airlock, it is exposed to the vacuum of space. As Dave frantically tries to close the hatch, there are about 15 seconds of film for which the soundtrack is utterly, spine-tinglingly silent. Only after he shuts the door and begins to fill the room with air does sound return.

The director, Stanley Kubrick, has often been praised for getting the physics right in *2001*, in contrast to the explosions and laser fire that accompany the space battles in *Star Wars, Star Trek*, and many other films.

However, it turns out Kubrick did not get it right after all, because there *are* sounds in space. As we will discuss in chapter 10, space might be more rarefied than anything we can produce in a laboratory here on Earth, but it is certainly not empty. In a typical part of the Milky Way, far from any stars, planets, or nebulas, every cubic yard of space contains about a million atoms. This is more than 10 million trillion times fewer atoms than in a cubic yard of air at sea level, but it is still not a vacuum.

Correspondingly, the pressure of the gas in space is extremely low. But because the pressure is not zero, the movements of stars, planets, and other celestial bodies through the cosmos will produce upward or downward variations in this pressure. And these pressure fluctuations will then travel through space as sound waves.

The noises that fill the Universe are certainly not sounds we're familiar with, and are at frequencies far, far below anything that humans are capable of hearing. But they are sounds all the same, no different or more exotic than other sounds we are unable to hear unaided, such as the ultrasonic screeches of bats or the deep rumblings of a whale song.

So what sorts of noises fill the Universe? Sounds far beyond anything our fragile ears can ever experience.

The ultimate sonic boom

A supernova is the catastrophic explosion with which a massive star ends its life. We saw in chapter 2 that supernovas are among the most luminous events in the Universe. Unsurprisingly, this intense light is also accompanied by a deafening boom.

When a supernova occurs, the outer layers of the star are blasted into space at enormous speeds. This results in a beautiful expanding bubble, known as a "supernova remnant," which can persist for many thousands of years before gradually fading from view.

A supernova remnant might appear fragile and delicate, but it represents a deafening wall of sound. To understand this, we first need to consider how sound moves. As I mentioned above, the speed of sound in the air around you is about a thousand feet per second. If you are standing a thousand feet away across an open field and I call out to you, there will be a delay of a second between when I speak and when you hear me: no surprises there.

However, there is another important implication of the speed of sound. Suppose you are in an underground subway station, waiting for a train. Usually you can tell when the train will be coming, perhaps a minute before you can see or hear it, because you can feel a rush of air coming out of the tunnel and onto the platform, pushed out by the train coming up behind it.

But why does this happen? Why doesn't the air just pile up in front of the train, like snow in a plow or dirt in a shovel? Shouldn't a train traveling through a long tunnel find the going increasingly harder as it sweeps up more and more air, until it grinds to a halt? Of course none of this happens. Before the train enters the tunnel, the tunnel is full of air. But as the train begins to enter, the air in the first part of the tunnel simply moves out of the way.

All this occurs because of sound waves. When the train first pushes into the tunnel, it indeed piles up air against it, just like a snowplow. This air now has a slightly higher pressure than normal, but it can immediately transmit this pressure variation deeper into the tunnel at the speed of sound. This keeps happening as the train keeps moving: The front face of the train tries to sweep up the air ahead of it, but the air responds sufficiently quickly that it can keep moving ahead and out of the way. What essentially occurs is that any air ahead of the train receives advance warning that the train is coming, and shifts farther down the tunnel before the train can hit it. The net result is that a gust of air is driven down the tunnel ahead of the train, which you can feel as a breeze on your face while standing on the platform.

But what happens if a train was to travel faster than the speed of sound, a thousand feet per second or around 760 miles per hour? While no trains travel at this speed, planes have been doing so for more than 60 years. And as we know, when a plane breaks the sound barrier, it results in a "sonic boom."

What happens is that the air in front of the plane receives no advance warning that the plane is coming, because sound waves can't move fast enough from the plane's current position to get to the next given parcel of air before the plane does.

Instead of a gentle rise in pressure, like the breeze being blown down a tunnel by an approaching train, there is an almost instant, huge, pressure jump, as completely undisturbed air in front of the plane suddenly finds itself squeezed against the plane's leading edge. Since sound is a variation in pressure, and since the larger the pressure change, the louder the sound, this results in a sharp, thunderous crash. Indeed, thunder itself is a naturally occurring sonic boom, produced when a bolt of lightning abruptly heats air to very high temperatures, causing it to expand at enormous, supersonic speeds.

Similar processes occur in space, the only difference being that the speed of sound is very different. The speed of sound in a gas depends primarily on its temperature. While space itself is very cold (see chapter 1), interstellar gas is generally quite hot, with a typical temperature of around 18,000°F. The corresponding sound speed in this gas is about 6 miles per second, or 22,000 miles per hour, about 30 times faster than the speed of sound in the Earth's atmosphere.

Some objects drift gently through the Galaxy at speeds well below this value. Just as for a train in a tunnel, any interstellar gas in their path simply shifts out of their way before they arrive. However, as we saw in chapter 5, many objects move far faster than the interstellar sound speed. Just as for a supersonic aircraft, this results in a deafening sonic boom. And the granddaddy of all sonic booms is that produced by a supernova.

The initial speed at which debris from a supernova explosion is driven outward into space can be as fast as 60 million miles per hour (which is around 10% of the speed of light!). This is almost 3,000 times faster than the speed of sound in space, and the pressure of interstellar gas correspondingly jumps up by an enormous factor when the supernova explosion slams into it.

How loud a boom does this make? To answer this question, we need to first consider how the loudness of sounds is measured. Loudness is a subjective phenomenon: What is a comfortable volume for one person might be way too loud for another, depending both on the quality of their hearing and their neurological perception of audio signals. So instead of talking directly about how loud a particular sound is, we need to step back slightly, and instead consider the "sound pressure" that a sound produces, that is, the strength of the air pressure fluctuations produced by the corresponding sound waves as they travel. A particular noise will have a specific sound pressure level, regardless of whether an individual person perceives that noise as loud or quiet.

We normally measure sound pressure levels in "decibels." Zero decibels is defined as a pressure fluctuation of 0.00000002% of normal air pressure, corresponding to the faintest sound a normal person can hear. Every factor of 10 in air pressure larger than this corresponds to an increase of 20 decibels. For example, when you hear your own voice in a normal conversation, the pressure fluctuations that your ear picks up are at levels around 0.000002% of air pressure. This is a hundred times larger variation in pressure than for zero decibels, so implies a volume of 40 decibels. The very loudest noises we might experience in everyday life, such as jackhammers, rock concerts, and jet engines, produce sound pressures of more than 100 decibels. And calculations suggest that the atomic bombs dropped on Hiroshima and Nagasaki reached around 250 decibels.

With this understanding in hand, we can use decibels to describe how loud a sound is in interstellar space, similarly adopting zero decibels to represent a pressure that deviates by 0.00000002% from the norm. The sonic boom from a nearby supernova produces a sudden increase in pressure of about 1,000,000,000%, which corresponds to more than 330 decibels!

If you could somehow float in space with your ears exposed as a supernova overran you, it's doubtful that you would hear anything at all. Even though the jump in pressure would be enormous, the overall pressure scale would be completely out of step with what those tiny hairs in your ear are capable of responding to. Your subjective experience would be that the supernova would be silent. But this would be a biased conclusion, based on the limits of our own perception. Have pity on some hypothetical alien species that lives between the stars, its ears attuned to the normal sounds of orbiting stars and drifting nebulas. To such a creature, a supernova would be utterly deafening, the loudest thing it or anything else could ever experience.

Hitting that bass note

A supernova is unbelievably loud but, like a thunderclap, it produces a single abrupt bang. However, the sounds of the cosmos are not just made up of individual crashes and cracks, but also of sustained notes and tones.

The difference between a high note and a low note comes back to the idea that a sound wave is a rapid oscillation of air pressure above and below the average. When listening to an orchestra perform a concert, all the sound waves travel at the same speed, and arrive at your ears in sync. However, there is a wide range in the *rate* of oscillation in pressure produced by different instruments. The deep notes of a double bass produce sound waves that vary back and forth in pressure only 50 times per second, while for a piccolo, a typical note will generate sound waves that oscillate more than 1,000 times every second. The number of pressure oscillations per second is known as the "frequency" of the sound wave, and is a direct measurement of the pitch of a musical note. Unlike loudness, which has a somewhat subjective nature, pitch has a direct mathematical foundation: What you will perceive as two notes an octave apart will differ in frequency by exactly a factor of two.

We can thus describe the pitch of sounds throughout the Universe in a very clear and well-defined manner. These sound waves might travel at very different speeds from those on Earth, and they might involve completely different scales of pressure level variations, but their rates of oscillations are something we can estimate from our astronomical measurements, and then put on a musical scale.

Astronomers' calculations suggest that the pitch of the Universe is very much a throaty bass or baritone rather than a soprano—most processes throughout the cosmos produce sounds that have much slower pressure oscillations, and hence much deeper notes, than we are used to hearing.

The deepest note yet identified belongs to a galaxy cluster, a conglomeration of several hundred galaxies and hot gas like those discussed in chapter 6. The particular cluster in question is Abell 426, often nicknamed the "Perseus Cluster" because of its location in that constellation.

Abell 426 is about 250 million light-years away. So at this great distance, how can we possibly know that it is putting out sounds, and how do we determine what note it is singing?

You might imagine that astronomers have built ultra-sensitive microphones, perhaps orbiting the Earth, which are capable of picking up the sounds from this cluster. We wish! At the speed of sound in interstellar or intergalactic space, the sound waves from Abell 426 would take more than 100 billion years to reach us, much longer than the age of the Universe. So we cannot directly hear Abell 426.

Instead, the approach that we must take might be considered analogous to that of a person with hearing loss living upstairs, while downstairs lives someone else who plays very loud music. The person on the top floor is unable to hear the music directly, but they can still tell that it is being played, because the salt and pepper shakers on their kitchen table vibrate in sync with the beat. Effectively, the person upstairs *sees* the sound waves rather than hears them.

In the same way, while we can never directly hear Abell 426's tune, we can see the pressure waves that it creates. For Abell 426, an additional challenge is that the actual vibrations happen far too slowly for us to watch them move back and forth. So to take the above analogy further, we don't get to see the salt and pepper shakers wobble in real time, but instead are reduced to taking a single photograph of the kitchen table.

With only a snapshot to go by, it would be challenging to realize that music was being played downstairs, let alone be able to state with any confidence what note was playing or how loud it was. But in the case of Abell 426, we have a crucial advantage: The gas that permeates the cluster (and in which individual galaxies are embedded) is incredibly hot, with a temperature in excess of 50,000,000°F. At this extreme level, this gas becomes incandescent, and radiates extremely energetic X-ray radiation.

In 2002, British astronomer Andy Fabian used NASA's

Chandra X-ray Observatory to make a detailed image of the X-rays produced by the hot gas in Abell 426. What these observations surprisingly revealed was a series of concentric ripples like those seen around a stone thrown into a pond. Through careful analysis of the data, Fabian and his colleagues were able to show that these ripples corresponded to places in the cluster where the density of the gas was slightly higher than the average. On the other hand, in the gaps between the ripples, they found that the gas density was slightly lower than average. Since a higher density means a higher pressure (and a lower density means a lower pressure), the conclusion was inescapable: These ripples are oscillations in pressure, a giant sound wave that thrums throughout this entire vast cluster.

Figuring out the pitch of the corresponding note is then relatively easy. We can calculate that the speed of sound in this 50,000,000°F gas is about 2.6 million miles per hour, and we can measure from the image that the spacing between each ripple is about 36,000 light-years. For a listener inside the cluster, with this sound wave rushing past them, we simply need to divide the speed of the wave by the spacing of the ripples to determine the rate at which the pressure wave oscillates, and hence which note is playing.

We can thus come to the remarkable conclusion that Abell 426, an unfathomably foreign object 250 million light-years away, is humming in B flat.

But this is a B flat unlike any you're familiar with: The sound waves in Abell 426 have an oscillation rate of once every 9 million years, which is 56 octaves below "middle C," or about 6,000 trillion times slower than the lowest note that the human ear can hear. Or to put it another way, you would need to add a further 635 keys to the left end of a piano keyboard in order to be able to play a note this low!

The deep bass note being sung by Abell 426 is nowhere near as loud as a supernova. The X-ray ripples seen in Abell 426 correspond to a variation in pressure of around 10%, which is a sound level of about 170 decibels, akin to holding your ear directly against the speaker during a major rock concert.

This is not nearly as loud as a supernova explosion, but a supernova is a single deafening crash. In contrast, Abell 426 has been playing a single sustained note for more than 2 billion years, and shows no sign of needing a breath any time soon.

The energy needed every second just to hold this note is staggering, a factor of a trillion trillion times larger than the combined output of every power station on Earth. What is the cosmic powerhouse that provides this energy, and why does it manifest itself as sound?

The culprit, as is so often the case for the most energetic events in the cosmos, is a black hole. At the center of Abell 426 is a large galaxy known as "NGC 1275." NGC 1275 in turn harbors a supermassive black hole at its core, with a mass of 400 million Suns. While this black hole is not directly visible, we can infer its presence through the enormous levels of light generated by the superheated mixture of stars and gas that continually spirals down into its throat.

Just as we saw in chapter 5 for the hypervelocity stars ejected by Sagittarius A* (the supermassive black hole at the center of the Milky Way), not everything that spirals close to a black hole ends up being consumed. At least some material always manages to escape, and is flung outward at high speed. Sagittarius A* has a very modest appetite: It takes 100,000 years to consume a mass equivalent to our solar system. Correspondingly, it only occasionally discards its scraps, as evidenced by the small number of hypervelocity stars that have been flung outward over millions of years.

In contrast, the much bigger black hole at the center of NGC 1275 is far more voracious, and gulps down the mass of our solar system every few weeks. Not surprisingly, the outflow of high-speed material that has experienced a near miss is also high. For reasons we as yet don't fully understand, this manifests itself in NGC 1275 and many other systems as two oppositely directed high-speed jets of material, which are blasted by the black hole over millions of light-years at a significant fraction of the speed of light.

In the case of NGC 1275, these twin jets need to force their way through all the gas of the large cluster in which the galaxy is enveloped. Like a garden hose running under water, the collision of the jets with the cluster gas generates a series of bubbles, which inflate under the jets' power, and then break off and float outward. As these bubbles expand, they shove the surrounding gas outward, setting up the pressure oscillations that then ring through the cluster in that deep B flat.

There are many galaxy clusters throughout the Universe, and there are huge numbers of supermassive black holes that generate a pair of jets of outflowing gas. But whether all these factors can combine together to make a single piercing note, and just what pitch that note is played at, hangs on a delicate combination of the rate at which the black hole feeds, the strength of the jets, and the size, density, and temperature of the surrounding cluster. Nevertheless, astronomers have now begun to identify other clusters that, like Abell 426, are good at holding a tune. The whole Universe appears to be rumbling with the harmonizing of the deepest throats imaginable.

The first sounds

At the present time, the Universe is filled with the deep humming of galaxy clusters, the sharp cracks of supernova explosions, and a myriad of other sounds. One way or another, all these sounds are produced by the varied motions and actions of stars, galaxies, black holes, and clusters. But these constituents of the cosmos have not always existed. We know that the Universe is 13.7 billion years old, and we know that there were times, very early on, when no stars or galaxies had yet formed.

So before the first supernova explosion, and before the first supermassive black hole, were the vast stretches of the Universe filled with nothing but silence? Or was there a cosmic song long before there were individual singers? What was the first sound in the Universe?

These sound like questions for philosophers. But incredibly, astronomers can answer them with considerable precision.

There is very strong evidence that space and time both began with an event known as the Big Bang, which from our current best estimates occurred 13.7 billion years ago. But despite its name, the Big Bang is thought to have been utterly silent. The distributions of matter and energy created in this sudden cataclysmic event were almost perfectly smooth— there were no oscillations in pressure that could correspond to any noise.

However, less than a trillion-trillionth of a second after the Big Bang, when the observable Universe had expanded to about the size of a beach ball, the cosmos had become decidedly lumpy. As time passed, and the Universe continued to expand, the denser clumps of material used their gravitational attraction to pull in more mass toward them. These clumps then grew in pressure as the gas in them became more tightly squeezed, forcing the gas to expand. As the clouds of gas expanded, their pressure dropped and their expansion slowed. Gravity then began to exert itself, and the process was repeated.

By less than a millisecond after the Big Bang, gas clouds over a whole range of sizes had begun collapsing and expanding, their pressure rising and falling as a consequence. Oscillations of pressure had been established—the Universe had found its voice!

These first sound waves were a little different from those we've talked about previously. Rather than traveling from point A to point B, like my voice sending sound through the air to your ears, these waves oscillated up and down in pressure without actually going anywhere. These are known as "standing waves," and are very similar to the stationary sound waves set up inside a flute or organ pipe.

The length of an organ pipe determines the tone of the sound it produces: The smallest organ pipes produce the highest notes. In an analogous way, the age of the observable Universe at these early times dictated the pitch of the primordial tune. When the Universe was very young, only clumps of matter that were relatively small, and for which the gas was able to expand and contract rapidly, had had enough time to complete one full cycle of pressure oscillations. Correspondingly, the cosmic choir was comprised only of sopranos. As the Universe aged, increasingly slower oscillations were completed, and correspondingly deeper notes were added to the chorus.

Furthermore, as time went on, the music became louder. This is because the overall level of clumpiness in the Universe increased as gravity began to exert its grip. As the clumps grew in size, the contrast between expansion and contraction of gas clouds was higher, and the pressure waves became stronger.

So what did the standing waves in the early Universe sound like? Just 10 years after the Big Bang, the dominant note in the Universe was F sharp (but 35 octaves lower than the lowest note a human ear can perceive), at a volume of 90 decibels (about as loud as standing next to a lawnmower). Over the next hundred thousand years, a whole new set of larger gas clouds was able to begin oscillating: More than 13 octaves of even deeper notes were added to the celestial pipe organ, with the volume increasing by a factor of 20.

At any moment in time, just as the largest possible gas cloud was completing its first cycle of collapse and expansion, there were other gas clouds, exactly half the size, which had completed two full cycles, and yet more clouds, half as large again, which had oscillated four times. As a consequence, the loudest note was accompanied by a whole set of fainter harmonics and overtones.

However, do not envision some pleasant-sounding (but thunderingly loud!) barbershop quartet. This set of harmonics was not the relatively pure timbre of a musical instrument, but a blurry blend of overlapping notes. The result, if you could hear it, would be a fuzzy hiss, steadily descending in pitch and gaining in volume as the Universe aged.

This celestial song lasted for 380,000 years, and then abruptly ceased, never to resume. What happened to mute this enormous cosmological pipe organ? And how do we know that these sounds ever happened, if they vanished billions of years ago?

As we discussed in chapter 1, at early times the Universe was a dense fog, because a ray of light was unable to travel even a short distance before colliding with an electron. It was throughout this period, known as the "pre-recombination era," that clumps of gas expanded and collapsed, producing these first sounds.

However, after 380,000 years, the Universe had cooled to a temperature of 4900°F, cold enough that protons and electrons could combine to form atoms. With this soup of free-floating electrons removed, the skies cleared, and the cosmos became transparent. This moment is known as "recombination."

This process silenced the Universe, because it changed the speed of sound. Before recombination, sound waves traveled through a gelatinous mix of light and matter, for which the speed of sound was about 60% of the speed of light, or about 400 million miles per hour. At this high sound speed, gas clouds were able to collapse and expand relatively quickly.

However, once matter and light went their separate ways, the speed of sound plummeted essentially to zero. At the moment of recombination all the sloshing of gas in and out immediately ceased, and the Universe became silent.

This sudden halt to the cosmic symphony, right at the time when the Universe opened itself up for view, means that we cannot see these sound waves as we can for the galaxy cluster Abell 426. So how do we know that they existed?

We know because although these sounds have long since faded, the final crescendo is forever frozen into the very fabric of the cosmos.

The moment of recombination left behind the cosmic microwave background (CMB), a faint, cold light that fills the Universe (see chapter 1). The CMB was discovered in the 1960s, and immediately became the object of detailed study by astronomers around the world. By the 1990s, precision observations were able to show that the glow from the CMB was not completely uniform, but that some parts of the sky were 0.001% warmer or cooler than others.

As measurements have continued to improve, these tiny variations (or "CMB anisotropies" as they are more formally known) have revealed a spectacularly detailed portrait of the Universe at that moment of recombination more than 13 billion years ago, just 380,000 years after the Big Bang. Because what these small temperature variations correspond to are individual clumps of gas, frozen in time in the middle of their pressure oscillations in or out. Those oscillations have now ceased their motion, but we can see them at their final positions. It is as if we have a photograph of the orchestra as it hits its final note: The conductor's arms are raised, and the performers can all be seen straining with effort as they play their instruments at their loudest volumes. But the sound itself is missing.

Astronomers have analyzed these temperature fluctuations in considerable detail, and have found that the CMB is not comprised of a random jumble of different-sized patches of hot and cold, but that regions of higher or lower temperature tend to have certain sizes. The clumps of hot and cold patches of temperature are almost all the same size, roughly twice the diameter of the full Moon. This implies that there is a clear fundamental tone imprinted onto the Universe (subsequent analysis has confirmed that this fundamental is accompanied by at least six higher harmonics). Specifically, we can state with considerable accuracy and confidence that the dominant note of the cosmos at recombination was almost exactly 54 octaves below middle C, at an ear-splitting volume of around 120 decibels. This is not quite as deep as the note sung by Abell 426, but is nonetheless remarkable: To play it, an organ would need a pipe more than 6 trillion miles long!

After recombination, the Universe continued to expand and cool, but did so in absolute silence. Over the next hundreds of millions of years, clumps of gas that happened to be near maximum contraction at recombination were able to continue collapsing under the influence of gravity, and eventually coalesced into the first stars and galaxies (see the discussion on Population III stars in chapters 3 and 6). As these various species gradually emerged from the emptiness, they not only restored light to the Universe but also sound, through noisemakers like the supernova explosions and galaxy clusters described earlier. And the Universe has never since stopped talking.

There is a final, startling connection between the strange harmonizing of the pre-recombination era and the hubbub that the cosmos experiences today.

As we can see directly from the CMB, the hottest gas clumps at recombination (that is, those that were just completing the compression part of their pressure oscillation at the moment the Universe became transparent) all had a particular size. The size that we see on the sky, about double the size of the full Moon, corresponded to a physical extent of 460,000 light-years at the time of recombination. However, over the 13.7 billion years since then, the Universe has expanded by more than a factor of 1,000. As a consequence, if these regions still existed, they would have been stretched so that they would now be 500 million light-years across.

In the early 1980s, astronomers began to measure the three-dimensional positions of hundreds of relatively nearby galaxies, and found that they are not scattered uniformly, but are clumped into complicated patterns (see chapter 4). The realization that the Universe is not totally chaotic but has a characteristic structure was a remarkable discovery.

But in 2005, when astronomers had expanded their catalogs of galaxy positions to many tens of thousands of objects, an even more incredible result emerged. Not only is the distribution of galaxies clumpy, but the size of these clumps is not random. How big is a typical clump of galaxies? Pretty close to 500 million light-years, the same size the hot clouds of gas from recombination would be if they had survived through to the present.

The conclusion is inescapable: These hot clouds *have* survived, but have now evolved into galaxies, stars, planets, and people. What we see all around us, and indeed are ourselves part of, is a fossil record of the oscillating sound waves from the earliest times in history, forever woven into the distribution of matter throughout the cosmos.

The first sounds in the Universe ceased long ago. The conductor and the musicians have departed the cosmic stage, taking their instruments with them. However, the performers have left behind their sheet music. By studying the CMB and the large-scale structure of the Universe, we can recover the first music ever played, music that was never meant to be heard.

Check Out Receipt

Free Library of Springfield Township
215-836-5300

Thursday, Jun 5 2014 12:49PM

Item: 31030004971727
Title: Extreme cosmos : a guided tour of the fas
test, brightest, hottest, heaviest, oldest, and
most amazing aspects of our universe
Material: Book
Due: 06/26/2014

Total items: 1

Please be sure to note our website: www.freelibr
aryofspringfieldtownship.org
Check out our many E-books & E-resources: Tumble
books, Zinio, Bookflix, and Freading

8

EXTREMES
OF ELECTRICITY
AND MAGNETISM

The modern world is completely reliant on electricity and magnetism. We use electricity to light our homes, watch television, and refrigerate our food. Huge industries are dedicated to generating it, and delivering it to the sockets in our walls. Magnets, meanwhile, are the key component of everyday items like credit cards, hard drives, microphones, and speakers.

Long before humanity learned to use them, electricity and magnetism inspired fear and awe in our ancestors. Lightning was something that even our earliest forebears knew to avoid, while the ancient Greeks knew that lodestones attracted iron.

In fact, although we think of electricity and magnetism as relatively modern conveniences, their application has had a surprisingly long history. Almost 2,500 years ago, the great Indian surgeon Sushruta recommended the use of magnets to remove iron arrowheads from wounds. And during the time of the Roman Empire, the physician Scribonius Largus wrote that placing an electric eel on a patient's forehead could cure a severe headache.

But it is only in the last 150–200 years that we have harnessed these mysterious and powerful phenomena, and have developed an understanding of their fundamental role in the Universe.

Two major concepts encapsulate our modern understanding. First, we know that electricity is made by electrons, the tiny elementary particles that were first discovered by the English physicist JJ Thompson in 1897. If a large number of electrons can be forced to flow together in a single direction (for example, along a wire), a current is generated and electricity is produced.

Second, we now know that electricity and magnetism are not separate phenomena, but are different facets of the same underlying, unified force called "electromagnetism." You can see this in the way a live wire deflects a nearby compass needle (electricity creating magnetism), or in the way that a spinning magnet in a dynamo will generate an electrical current (magnetism making electricity).

Indeed, we now know that light itself is an "electromagnetic wave," in which an electric field generates a magnetic field, which then produces a new electric field, which then creates another magnetic field, and so on.

Electromagnetism does more than simply power our television and keep the utility bill attached to the refrigerator door. It is one of the fundamental forces that control the cosmos, and naturally occurring electricity and magnetism are everywhere we look.

However, before we can discuss the properties of the

electromagnetic Universe, we need to consider how we measure and describe these forces. Everyone is familiar with voltage and current. For example, the computer on which I am writing these words draws a voltage of 240 volts at a current of 1.5 amps. But let's think a little more about what these terms mean.

Current simply measures the rate at which electrons flow along a wire. A current of 1 amp is produced when approximately 6,241,509,600,000,000,000 electrons pass a given point every second.

Voltage is a slightly more abstract concept: It measures the strength of the force with which electrons are pushed along the wire. So a 1-amp current with a voltage of 4 volts can produce twice as much electrical power as the same current with a voltage of 2 volts: The electrons flow along the circuit at the same rate in both cases, but in the first case the electrons are pushed twice as hard.

Imagine yourself and a large group of friends standing on top of a hill, next to a box filled with dozens of solid metal balls of different masses. Another group of friends stands at the bottom of the hill, and it's their job to catch the balls that the first group rolls down.

Suppose the group at the top of the hill now rolls several balls simultaneously down the slope, and that everyone has agreed to use balls of the same mass. Think of these balls as electrons, and their journey down the hill as an electric current. In this case, the strength of the current (in amps) corresponds to the number of balls that are released, while the voltage is represented by the force with which each ball hits the hand of the person catching it.

So suppose that you and your friends first roll 10 small balls down the hill all at once. After these 10 balls have been caught at the bottom, you do this again, but with 20 small balls.

To an observer standing halfway down the hill, the number of balls that pass them has doubled, so the current has doubled. However, in both cases the force that a person at the bottom experiences as they catch the ball is the same, so the voltage has not changed. (The only difference is that twice as many people at the bottom are needed to catch the extra balls.)

The group at the top now takes 10 new balls out of their box, with each new ball having twice the mass of one of the first set of balls. When these heavier balls roll down the hill, the number of balls is the same as the first time, so the current is the same (and the balls roll at the same velocity as before, since gravity causes all objects, regardless of mass, to fall at the same speed). But because these new balls are twice as heavy, they hit the hands of the catchers with twice the force of the small balls, so that the voltage has doubled.

This is a simple analogy, which you should not use as the basis for your understanding of electromagnetic theory! (For example, a higher voltage does not actually correspond to heavier electrons; all electrons are the same mass.) Nevertheless, it can provide a basic picture with which we can compare our everyday experiences of electricity to what occurs elsewhere in the Universe.

That's electricity—what about magnetism? For magnets, there are several different measurements we can make. The most obvious might be how strong a pull a magnet exerts on things around it. For example, a magnet that can just lift an iron nail weighing half an ounce has a "pull force" of half an ounce.

However, comparing the pull force of two different magnets can be misleading, because it depends critically on the shape of the magnet and the direction in which you point it. And if we cannot actually touch the magnet to an object to test its strength (as is the case for a magnet in space that we want to study), then the pull force is not an especially meaningful concept.

An alternative approach to measuring magnets is based on the fact that magnetic fields are generated by electric currents (even a fridge magnet has tiny circulating currents in it at atomic scales).

According to this approach, the intrinsic strength of a magnet—known as its "magnetic moment"—can be determined by multiplying the strength of the magnet's interior electric current by the area of the loop that this current makes as it completes its circuit.

So a magnet with a large magnetic moment will have a strong internal electric current, a large size, or both. Generally speaking, the pull force increases with the square of the magnetic moment: If you double the magnetic moment, the magnet can now lift a nail four times as heavy. However, the key point is that using the magnetic moment allows us to compare different magnets, regardless of their shape or orientation, and whether they are out in space or sitting on the table in front of us.

And finally, we need to bear in mind that the attractive or repulsive force of a magnet drops off as the magnet moves farther way. Two magnets with the same pull force or same magnetic moment will have different attractive powers if one is more distant than the other. We can thus also choose to describe the *magnetic field* at a particular location, rather than worry about the specifics of the properties of the magnet and its distance from us. If we don't know exactly how a magnet works or how far away it is, then the magnetic field is often the only thing we can easily measure.

For example, there is a giant natural magnet in the Earth's molten metallic core, which causes all compasses to point north. It makes no sense to describe the pull force of this magnet for several reasons: For example, it has no solid surface that one could touch to an object before trying to lift it. We also cannot directly determine the magnetic moment of the Earth, since we have no way of measuring electrical currents thousands of miles beneath our feet. However, the one thing we can quickly and easily determine is the magnetic field on the Earth's surface, simply by observing the twist that this magnetic field exerts on a compass needle as it aligns itself with north. Magnetic fields are measured in "gauss," and the magnetic field on the surface of the Earth is typically about half a gauss.

The higher the field, the stronger the twist on the compass needle. For example, the magnetic field at the surface of a fridge magnet is about 50 gauss, a hundred times stronger than the magnetic field of the Earth. A fridge magnet held next to a compass will therefore twist the compass needle a hundred times harder than the Earth can, even though the total magnetic moment of the Earth is far larger than that of a fridge magnet.

With these concepts in hand, we can now proceed to examine the spectacular extremes of electricity and magnetism that the Universe can generate.

Galactic lodestones

Planets, stars, and even entire galaxies are all magnetic.

It might surprise you to hear that there are magnets in space. But recall that a magnet is produced whenever you have an electric circuit—that is, a flow of electrons around a closed loop. Most heavenly bodies contain electrons, and these bodies virtually always rotate. All these objects therefore have circulating currents inside them, and so act as magnets.

I have already mentioned the Earth's magnetic field, which at its surface has a strength of about 0.5 gauss. This field extends up into the atmosphere and for tens of thousands of miles out into space. The Earth's magnetism is vital for the migration of birds (some species of which seem to be able to see magnetic fields directly). It produces the glowing auroras seen above the North and South poles. And it shields us from the torrent of harmful particles with which the Sun continuously bombards the Earth.

The Sun too has a magnetic field, and over most of the surface of the Sun the field is just a few times larger than the Earth's. However, sunspots mark small regions of intense magnetism, where the field is more than 1,000 times stronger than average. Occasionally these strong magnetic fields abruptly change shape and readjust. An explosive release of energy results from this rearrangement, which is what produces dramatic events like solar flares. A spectacular example of such a flare was seen on March 30, 2010. Within a matter of minutes, a rapidly expanding magnetic ring of hot gas was thrust more than 120,000 miles above the Sun's surface, eventually breaking open and spewing solar material into space.

We often see the equivalents of solar flares on other stars, demonstrating that they too are magnetic. Potentially the most spectacular example of this occurred on the evening of January 5, 2009, in a faint red dwarf star (see chapters 3 and 6) known as "YZ Canis Minoris." In the space of just 10 minutes, YZ Canis Minoris abruptly brightened by a factor of 200, before slowly fading back to normal over the next 10 hours or so. The magnetic fields responsible for this sort of dramatic activity on the surface of these "flare stars" can be 3,000–4,000 gauss.

A special category of stars, known as "Ap stars," is even more magnetic. In 1960, American Horace Babcock astounded the astronomical community when he presented measurements of the surface magnetic field for the Ap star HD 215441, with a strength of more than 34,000 gauss! Many other Ap stars with high magnetic fields have since been identified, but HD 215441 still holds the record for the most magnetic normal star ever found. Scientists now generally refer to HD 215441 as "Babcock's Star" to honor this discovery.

Now 34,000 gauss might sound like a strong magnet, but it's a magnetic field that most people can experience without any ill effects. Indeed, you or someone you know has probably had an MRI scan, a medical scanning technique that routinely uses magnetic fields of around 80,000 gauss.

Physicists have built much stronger magnets than this for their laboratory experiments. The world record belongs to the Tesla Hybrid Magnet at Florida State University, which can generate a magnetic field of 450,000 gauss. The Tesla Hybrid is not a giant lump of iron, but an electromagnet, meaning that electric currents generate the magnetism, and can be switched on and off. To function, the Tesla Magnet needs to be kept at a temperature of −456°F (almost as cold as the Boomerang Nebula discussed in chapter 1) and requires 33 megawatts of electricity.

Even higher magnetic fields can be created for brief periods. The Multishot Magnet in Los Alamos produces brief, extreme magnetic pulses. Its current record is 889,000 gauss (which it was able to sustain for only 15 milliseconds), and there are plans to push the Multishot Magnet past the 1,000,000 gauss barrier in the near future. Finally, scientists in Russia have designed an experiment called "MC-1" in which they use 370 pounds of high explosives to achieve a magnetic field of 28 million gauss for a few microseconds. However, this is inevitably a single-use magnet—there is not much left of the equipment afterward!

MC-1 briefly produced a magnetic field almost 1,000 times stronger than Babcock's Star. So perhaps it would seem that at least for magnets, humans can outdo the rest of the cosmos. Not quite.

What would happen if Babcock's Star were to somehow shrink? Shrinking a star intensifies its magnetism considerably—the magnetic field on the surface will increase as the inverse square of the star's diameter, so if Babcock's Star were to collapse to half its size, the magnetic field would increase by a factor of four.

This is not a fanciful scenario, because when stars exhaust their supply of fuel, they do indeed shrink. As we saw in chapter 1, in about 5 billion years the Sun will use up all the hydrogen and helium in its core. Without the heat and pressure from nuclear fusion, its central regions will then collapse under their own gravity, while the outer layers will flow off into space to form a glowing planetary nebula. The collapsing core of the Sun will continue to shrink until it is only around 9,000–12,000 miles across—not much more than the diameter of the Earth. Shrunk to a much smaller size, this newly formed white dwarf will have a vastly more intense magnetic field on its surface than the Sun could ever hope to achieve.

How magnetic can white dwarfs get? The current record goes to the white dwarf "PG 1031+234" in the constellation Leo, whose surface magnetic field is in some places as high as 1 billion gauss! This is almost 40 times as intense as the magnetic field reached by the MC-1 experiment in Russia. And while MC-1 could only maintain its magnetic field for a few microseconds, incredibly PG 1031+234 has been comfortably maintaining its much more intense magnetism for many millions of years.

Can we get even more magnetic than this? Oh yes.

In chapters 3 and 5, we talked about neutron stars. Just as a white dwarf is the small, collapsed core left behind by a star like the Sun, a neutron star is the even smaller, denser remnant that remains after a much more massive star ends its life in a supernova.

Neutron stars are extreme in almost every way: They are only about 15 miles across, and can spin many hundreds of times per second (see chapter 3). Since they result from the collapse of a very large star into a very small one, this process intensifies their magnetism by a spectacular degree.

As I explained in chapter 3, many neutron stars manifest themselves as pulsars, emitting a pulse of radio waves once per rotation as their lighthouse beam swings around the sky. The speed at which a pulsar spins is not quite constant; careful measurements show that these stars are all slowing down at a minuscule rate. We believe that this "spin down" is simply due to the fact that pulsars are spinning, superstrong magnets. In the same way that a rotating magnet in a dynamo can drive an electrical current, a spinning pulsar produces electrical currents in its surroundings, and uses up some of its spin energy in the process. We can therefore calculate the surface magnetic field of any pulsar simply by knowing how fast it is spinning and at what rate it is slowing down.

The result that one quickly establishes through this technique is that even a "garden variety" pulsar has a magnetic field so large that it borders on incomprehensible. The most magnetic pulsar yet discovered, known as "PSR J1847-0130," has a surface magnetic field of 100 trillion gauss! This is 100,000 times more magnetic than the white dwarf PG 1031+234, or 3 billion times more magnetic than Babcock's Star.

But it doesn't stop there. The ultimate title for strongest magnets in the Universe goes to an extremely rare species of neutron star known as "magnetars" (there are 2,000 known pulsars, but only about 25 known magnetars).

The discovery of magnetars is a marvelous detective story.

It began with a bang at 3:51 p.m. Greenwich Mean Time on Monday, March 5, 1979, when a dramatic pulse of energetic gamma rays suddenly slammed into every space probe in the solar system. (The Earth's atmosphere protected those of us on the ground from this harmful radiation.)

This burst of energy overwhelmed different spacecraft at slightly different times. The relative time delays allowed astronomers to triangulate the location on the sky of this strange event: It originated in the constellation of Dorado, and probably came from an object in the Large Magellanic Cloud (our neighboring galaxy, 170,000 light-years away).

For many years this strange blast of gamma rays remained a mystery. (As a teenager, I even remember reading a book about unsolved problems in astronomy, in which "the March 1979 event" featured heavily.) All sorts of weird and wonderful ideas were put forward to explain it, but the one that eventually turned out to be right was a theory proposed almost simultaneously in mid-1992 by Vladimir Usov, by Bohdan Paczyński, and jointly by Robert Duncan and Chris Thompson. What all three groups suggested was that a neutron star with a ridiculously strong magnetic field, even stronger than the most magnetic pulsar known, could potentially have produced this strange flash of gamma rays. Duncan and Thompson in particular were so confident of their theory that in the second sentence of their published paper, they announced that these hypothetical super-magnetized stars should henceforth be called "magnetars." (The name has certainly stuck, and is now even defined in the *Oxford English Dictionary* as "a neutron star with a much stronger magnetic field than ordinary neutron stars.")

Magnetars remained a wonderful but unproven idea until November 1996, when a team led by Greek-American astronomer Chryssa Kouveliotou made a detailed study of an unusual star known as "SGR 1806-20." What Kouveliotou and her colleagues discovered was that SGR 1806-20 was a spinning, pulsing neutron star. But was SGR 1806-20 just another pulsar? Absolutely not.

First, SGR 1806-20 was spinning much more slowly than a typical pulsar: Its regular pulses showed that it took 7.5 seconds to turn once on its axis. This is more than 10,000 times more rapid than the Earth's rotation, but is positively sluggish compared to most pulsars (see chapter 3). Second, SGR 1806-20 was slowing down in its rotation at a rate 100 times faster than any pulsar. The clear conclusion was that SGR 1806-20 is a spinning magnet just like a pulsar (and was probably born spinning rapidly, also like a pulsar). However, its much greater level of spin down requires a much stronger magnetic field. How strong? The surface magnetic field of SGR 1806-20 is more than 1,000 trillion gauss!

Magnetars were no longer just a clever idea. They were real. About 20 more magnetars have since been identified, all with surface magnetic fields of hundreds of trillions of gauss. But it is SGR 1806-20 that still easily holds the title of strongest known magnet in the Universe.

Needless to say, it is hard to imagine a magnet this strong. This is so intense that even if SGR 1806-20 were 600,000 miles away (almost three times the distance to the Moon), its magnetism would overwhelm that of the Earth, making compass navigation impossible. At a distance of 60,000 miles, it would wipe the data from every credit card and hard drive on the planet. And at 10,000 miles, the magnetic field of SGR 1806-20 would be fatal, because it would now be so intense that it would disrupt the electrical nerve impulses that make your heart beat.

Fortunately, SGR 1806-20 is more than 30,000 light-years away, and so does not produce these effects! However, the magnetism of SGR 1806-20 is so intense that even at this immense distance, it can directly affect the Earth in other small but measurable ways.

As I mentioned above, the Sun produces solar flares when

the magnetic field on the Sun's surface suddenly changes shape and snaps into a new configuration. About once per century, the same thing happens on a magnetar. But in this case, the strength of the magnet is much, much higher, and the results are far more dramatic, generating a tremendous explosion of radiation and energy. Astronomers now believe that the event of March 5, 1979, was just such a "giant flare," from a magnetar now known as "SGR 0526-66." And on August 27, 1998, another giant flare walloped the solar system, this time from a magnetar called "SGR 1900+14."

But the biggest flare was yet to come.

On December 27, 2004, it was the turn of SGR 1806-20, the most magnetic star of all, to show us what it could do. At 9:30 p.m. Greenwich Mean Time, hundreds of satellites and spacecraft, both orbiting the Earth and spread throughout the solar system, were all completely overwhelmed for 0.6 of a second by a flare of gamma rays, with an intensity 10,000 times the level of the first giant flare of 1979! Other than the Sun, this was easily the strongest blast of radiation ever recorded from any celestial body. The flare was so strong that one satellite even detected a second burst of gamma rays, 2.6 seconds after the initial flash, due to the echo of the signal off the Moon.

I was lucky enough to be one of the astronomers in the box seat when this incredible event occurred, and led a team that used a fleet of radio telescopes spread over the globe to study the amazing aftermath of this spectacular magnetically powered explosion. What we saw afterward was that about 10 billion billion tons of material (about the mass of Pluto) had been blasted off the surface of the neutron star at 50% of the speed of light. Several years later, the glowing cloud of debris is still just detectable by our most powerful telescopes as it continues to expand and fade.

The energetics of the giant flare from SGR 1806-20 are staggering. In just a fraction of a second, SGR 1806-20 put out more heat and more light than the Sun does in 150,000 years. Or to put it another way, for 0.6 of a second, SGR 1806-20 outshone the entire Milky Way by a factor of more than 1,000!

Apart from its extreme intensity, what was remarkable about this event is that it had a measurable impact on the Earth. We normally don't duck for cover when a distant supernova explodes, or brace for impact when we see two galaxies collide. But in the case of SGR 1806-20, this little magnetar actually reached across tens of thousands of light-years and tapped the Earth on the shoulder. The giant gamma-ray flare from SGR 1806-20 created a massive disturbance in the Earth's ionosphere, abruptly affected the radio communication system used by the US Navy's fleet of submarines, and even slightly altered the strength of the Earth's magnetic field.

Things returned to normal after less than an hour, and no lasting damage was done. Nevertheless, the episode was a stark reminder of the colossal power and energy of the cosmos, and demonstrated that we cannot always be detached, uninvolved observers. Even from a distance of 30,000 light-years, the Universe's strongest magnet can make itself felt here on Earth.

High voltage rock and roll

High voltages are all around us. We've all experienced the jolt of static electricity when turning a doorknob or getting into a car on a dry day. Such shocks are harmless, because they transmit only a tiny amount of current. But the voltages involved can be surprisingly large, usually many thousands of volts.

The Van de Graaff generators on display at science museums throughout the world can go much higher, easily reaching voltages of over 1,000,000 volts. Putting your hand on one can charge you up with so much static electricity that your hair stands on end. But if used properly, there is again no risk of electrocution or injury, because the currents are tiny.

The highest voltage machine ever built is the "Holifield Radioactive Ion Beam Facility" (HRIBF), a giant experimental atomic physics facility located in Oak Ridge, Tennessee. In regular operation, the HRIBF uses a voltage of around 25 million volts. In May 1979, when the equipment was being tested, it reached a voltage of 32 million volts.

By nature's standards, this is not very impressive. Even without needing to leave Earth, we can outdo the HRIBF: A typical lightning bolt carries 100 million volts, with some strikes observed to exceed a billion volts (generally speaking, the longer the lightning bolt, the higher the voltage).

But in space, the voltages are much larger.

The ultimate pocket dynamos are pulsars, the spinning celestial magnets that we discussed above.

To understand how pulsars generate enormous voltages, let's look back for a moment to the work of the great English physicist Michael Faraday. In 1831, Faraday invented an experiment called a "homopolar generator." In this device, a copper disc spins while magnets sit both above and below its rim, generating a voltage between the center of the disc and its edge. This landmark invention demonstrated that mechanical motion could be converted into electricity, and was the fore-runner of the electrical generators that now power our lives.

The surface of a pulsar is a conducting material like a copper disc, while as we have seen, the magnetic field at this surface is stupendously large. As the pulsar furiously spins, its surface acts as a naturally occurring homopolar generator, but with voltages unimaginably larger than anything Mr. Faraday could have envisioned.

The voltage that a pulsar generates depends on two things: the strength of the magnetic field, and the square of the rate at which the star rotates. Because both these factors contribute, some of the spectacular neutron stars that we met earlier do not have as high a voltage as you might think. SGR 1806-20, the magnetar with a surface magnetic field of 1,000 trillion gauss, spins too slowly to generate an extreme voltage. And PSR J1748-2446ad, the pulsar in chapter 3 that spins 716 times per second, has a magnetic field that is too weak. To find the highest voltage, we need to look to PSR J0537-6910, a young pulsar that spins 62 times per second and has a surface magnetic field of 925 billion gauss. With both rapid rotation and strong magnetism, PSR J0537-6910 generates a voltage of around 38,000 trillion volts! A lightning bolt with this voltage would need to be about 2.5 billion miles long, or about the distance from Earth to Neptune.

But for voltages, pulsars are surpassed by the masters of the extreme: supermassive black holes.

As we have seen in chapters 5, 6, and 7, enormous black holes sit at the centers of most large galaxies, including the Milky Way. A typical supermassive black hole weighs 100 million times the mass of the Sun and is about 100 million miles across. In chapter 5, we discussed how superheated gas steadily spirals into such a black hole, like water down a drain. Not surprisingly, like almost everything else in the Universe, these clouds of gas are expected to be magnetic.

But a critical extra part of the story is that we expect the black hole itself to also be spinning; a typical supermassive black hole rotates on its axis about once every two hours. In 1977, astronomers Roger Blandford and Roman Znajek published a landmark calculation, in which they showed that if a spinning black hole is steadily fed by hot, magnetic gas, it forms a giant electrical circuit. As the black hole rotates, currents flow out of the equator of the hole, through the gas surrounding it, and then back into the black hole's pole. Just like a giant battery, a spinning black hole can generate electricity. And because of the huge mass involved, the voltage is staggering, around 10 million trillion volts.

People normally associate black holes with strong gravity (see chapter 9): Anything that strays too close runs the risk of being sucked in, never to return. But next time you decide to visit a spinning supermassive black hole, you'll be making a serious mistake if you worry only about watching out for gravity. Even if you're well outside the range where the black hole's gravitation becomes a problem, the extreme voltage of the hole and its surroundings will ensure that your visit will be an electrifying one.

The longest lightning bolts of all

The few thousand volts that jolt us when we get a static shock from a car door are quite harmless, but currents are a very different story. Even a current flow of just 0.1 of an amp can be lethal. The Universe routinely generates currents far beyond this—the cosmos is not a safe place for fragile humans!

Here on Earth, a lightning bolt might typically carry a current of 20,000–50,000 amps, while the auroras above the Earth's North and South poles correspond to current flows of around 1,000,000 amps. This might sound impressive, but is well below the strongest current ever seen on our home planet. That landmark was set at Sandia National Laboratories in Albuquerque, New Mexico, by a piece of equipment known as the "Z Machine." The Z Machine creates conditions of extreme temperature and pressure by firing brief but massive electrical pulses through tiny tungsten wires. In late 2007, the Z Machine set a new record, hitting a current of 26 million amps, albeit only for about a tenth of a microsecond. In the future, the Z Machine and similar machines should be able to power currents up to 70 million amps.

However, once we move beyond the Earth, the strength of the currents outstrips what even our most powerful experiments can generate. On the Sun, active regions such as sunspots are sites of intense magnetic fields and current flow. A typical sunspot lasts for a week and carries a current of around a trillion amps. This is around 40,000 times the current produced by the Z Machine, and persists for about 6 trillion times longer.

We saw above that pulsars generate colossal voltages, and with them come enormous currents also. The pulsar PSR J0537-6910, with its furious spin rate and its voltage of 38,000 trillion volts, generates a current of about a 1,000 trillion amps. The resulting energy output is equivalent to detonating

10,000,000,000,000,000 one-megaton atomic bombs every second! Not surprisingly, this vast electrical discharge puts on quite a show. Many of the most energetic pulsars are embedded in "supersized" versions of the Earth's auroras, resulting in spectacular glowing nebulas several light-years across.

But yet again, the last word must go to supermassive black holes. Black holes do not suck down everything that comes near them. The deep bass note in Abell 426 that we discussed in chapter 7 is powered by matter that swirls down toward the maw of a supermassive black hole, but is then flung outward in a pair of oppositely directed narrow jets, traveling at light speed.

While in the case of Abell 426 these jets produce the gas vibrations associated with a deep deafening note, for many other supermassive black holes these jets travel unimpeded for up to a million light-years, before abruptly slamming into unsuspecting clouds of intergalactic gas. This collision heats these gas clouds, and causes them to shine brightly in the light of radio waves.

The resulting objects are known as "radio galaxies," and are some of the most powerful and energetic beasts in the Universe. The nearest one, "Centaurus A," stretches about 16 times the diameter of the full Moon across the sky, even though it is more than 12 million light-years away.

Electric current simply measures the rate at which electrons flow along a wire. The jets that power radio galaxies are the ultimate celestial power lines, with huge numbers of electrons flowing along them. The enormous currents that result play a major role in keeping these jets so long and straight—if the jets were not electrified, they would flop and bend and not be anywhere near as spectacular as they appear.

Just how much electricity is involved? The jets in radio galaxies carry the highest observed currents in the Universe, typically at the level of 1 million trillion amps. Their power output is so large that in a single millisecond, a radio galaxy provides enough electricity to cover all of humanity's energy needs for the next 20 trillion years! If we ever exhaust our fuel and power supplies here on Earth, radio galaxies can assuredly cover the shortfall.

9

EXTREMES
OF GRAVITY

As Isaac Newton brilliantly realized, an apple falls to the ground for the same reason that the Moon orbits the Earth: gravity.

Gravity is the universal force. Every particle of matter in the cosmos exerts a gravitational pull on every other particle. While the only gravity you can "feel" right now is that of the Earth pulling you down to the ground, the people and objects around you are all also tugging on you with their own minute gravitational forces, as is every planet, star, and galaxy in the entire Universe.

While gravity is universal, it is surprisingly weak. This can be simply demonstrated by touching a small magnet to a paper clip and lifting it up in the air. Magnetism is so much stronger a force than gravity that a tiny magnet can overcome the gravitational pull of the entire planet Earth. The key difference is that magnetism generally does not extend over very large distances, while gravity is everywhere. Gravity cannot be shielded from, hidden from, or eliminated. Everything in the Universe experiences the gravitational attraction of everything else.

This fact is clearly apparent from the structure of the cosmos. Planets orbit stars. Stars orbit around the centers of galaxies. Galaxies trace complicated orbits within galaxy clusters. Almost all the processes we see in the Universe are ultimately initiated and maintained by gravity.

However, depending on where you are, gravity manifests itself in different ways.

If a massive body holds you to its surface with its gravity, as the Earth does to all of us, then you perceive the pull of gravity as your weight. However, it's important to understand that it's not possible to feel gravity directly. What you feel as weight is not the Earth's gravity pulling down, but the chair on which you are sitting or the ground on which you are standing pushing up. The Earth is trying to drag you down toward its center, but the chair or floor exerts a force of its own, of the same strength but in the opposite direction, keeping you in place. It's this resistance to gravity's pull that you feel as weight, not gravity itself.

The weight you experience when you're standing on a planet or other object in space depends on the object's mass and on its diameter. For example, the Moon has one-eightieth the mass of the Earth, and 27% of its diameter. Combining these two effects, the weight you experience when standing on the Moon is about one-sixth that which you normally feel on Earth.

But what if you are not standing on the surface of a planet or other massive body, but are simply falling or drifting through space, following gravity's pull? Astronauts on the International

Space Station are in exactly this situation as they orbit the Earth every 90 minutes. We've all seen the video footage of life on the space station: Pencils, equipment, and even the astronauts themselves float around the capsule unless they are tied down. In everyday speech, we use the phrase "zero gravity" to describe this situation. But from a physics perspective, this is not correct.

You might think astronauts float because they are a long way from the Earth, and gravity is weaker. The astronauts are about 200 miles above the Earth, so indeed the gravitational pull they experience is reduced compared to its strength at the Earth's surface. But this is a modest effect: At these heights, the Earth's pull due to gravity is only about 10% lower than what we experience at sea level. The space station is certainly not far enough away from the Earth for the gravity it experiences to drop to a level small enough to be considered "zero gravity."

So what's going on?

Astronauts in the space station are fully in the grip of gravity. We know this because they are orbiting the Earth. Clearly it is gravity that causes them to orbit—otherwise the space station and everyone in it would drift away from the Earth and off into deep space. But because the astronauts are in free fall rather than standing on the surface of the Earth, there is nothing to resist the gravitational pull, nothing to push back and keep them in place. Without this resisting force (referred to in physics as the "normal force"), the sensation of weight disappears, and everything floats. Astronauts very much experience "weightlessness," but not zero gravity.

So weightlessness has nothing to do with how high up you go or how fast you orbit. All you need to do to become weightless is to put yourself in free fall. Jump off a high diving board, and for those couple of seconds as you plunge toward the water, you are as weightless as any astronaut.

In order to consider extremes of gravity throughout the Universe, we need to decide how to describe the gravity of different objects. If something has a solid surface that can support you, then you can directly measure the object's gravity by the weight you experience when standing on it. But if you are orbiting an object rather than standing on its surface, or if the object is a gas cloud or a galaxy that doesn't have a surface in any case, then you will be weightless no matter how strong the object's gravitational pull. In this case, we can describe the strength of gravity by how quickly something would fall downward if you dropped it. For example, if you drop a stone into a deep well on Earth, it will fall much faster than if you do so on the Moon. If you are a light-year away from a large cloud of gas, it still makes sense to imagine the experiment of dropping a stone down toward the cloud, and to see how rapidly it falls.

It's important to note that this hypothetical situation only makes sense if you are somehow hovering stationary relative to the cloud, rather than moving or orbiting. If an astronaut in the space station leans out of the hatch and lets go of a stone, the stone will not fall to the surface of the Earth, but from the astronaut's perspective will appear to float in space, remaining exactly where it is released. This is because before the astronaut released the stone, both the astronaut and the stone were already orbiting the Earth. Once the stone is set free, nothing changes—the stone and astronaut continue to follow parallel tracks around the Earth, and from the astronaut's perspective the stone stays where it was. Or to put it another way, the astronaut was already falling, and whether she holds on to the stone or lets it go, the stone falls at the same rate.

But now imagine if the astronaut could use a jet pack to stop in her orbit, and so hold her position at one place above the Earth. A stone released from this vantage point would fall downward, just like a stone dropped down a well. The situation is contrived, but it emphasizes the difficulty of experiencing or measuring gravity unless you have a place to stand. In the wider Universe, there are few safe or sensible spots to stand. So to consider the extremes of gravity, we must bring our jet pack and a bagful of stones on our journey.

A weighty issue

The strength of the Earth's gravity near its surface can be measured by the acceleration an object experiences as it falls. Anyone who has bungee jumped or skydived knows exactly what this feels like.

To put this in mathematical terms, we can say that the acceleration due to the Earth's gravity is 32 feet per second per second, or about 22 miles per hour per second. What this means is that if you fall to the ground from a large height, then as every second passes, your downward velocity (ignoring air resistance) will increase by 22 miles per hour. One second after you begin falling, you will be moving at 22 miles per hour. After another second has passed, you will be falling at 44 miles per hour. And a second after that, you'll have hit 66 miles per hour, and so on.

Or to put it another way, acceleration due to gravity of 22 miles per hour per second corresponds to the normal sensation of weight that we are used to experiencing on Earth. If you stand on the surface of a solid body with a higher gravitational acceleration than this, you will feel heavier.

Using this point of reference, we're ready to ask where the strongest gravitational forces in the cosmos occur. The strength of an object's gravitational pull depends both on that body's mass and on the inverse square of your distance from it. So it's reasonable to expect that the strongest gravitational forces in the Universe will result from getting close to very massive but very small objects.

Earlier in this book, we encountered three types of unusually dense objects: white dwarfs, neutron stars, and black holes. All have unimaginable extremes of gravity.

Let's start with white dwarfs, the dense, hot cores that remain after stars like the Sun use up all their nuclear fuel. A typical white dwarf might have the same diameter as the Earth, but a mass 300,000 times larger. That means that gravity at the surface of a white dwarf is 300,000 times stronger than the Earth's. Instead of the usual acceleration of 22 miles per hour per second that we experience when we're falling, the gravitational acceleration of a white dwarf is around 6.5 *million* miles per hour per second. If you dived off a 30-foot platform into a swimming pool on a white dwarf, it would take you about 8 milliseconds to hit the water, compared to about 1.4 seconds on Earth.

A neutron star is even smaller and more massive than a white dwarf, so its gravity is even stronger. An average neutron star has 40% more mass than the Sun, but is just 15 miles across. So the acceleration due to gravity on the surface of a neutron star is an enormous 3 trillion (that's 3 million million) miles per hour per second. Your dive into the swimming pool would now last just four-millionths of a second! The pull of gravity would be so intense that to move off the surface of a neutron star by 1 inch would require 38,000 times more effort than climbing Mount Everest.

And of course, black holes are even more compact and more massive than neutron stars, so their gravity must be even more intense. But before considering the case of black holes, note that black holes don't have an actual surface that you could stand on. The closest equivalent is an imaginary surface known as the "event horizon," which marks the point of no return. If you are just outside the event horizon, you experience extreme gravity from the black hole, but you can still get away if you accelerate extremely hard in the opposite direction. But once you cross the event horizon, there is no escape: Even if you could travel at the speed of light, you still would not be moving fast enough to get away from the black hole's clutches. So while we can't talk about standing on the surface of a black hole, we can characterize a black hole's gravity by considering the situation in which you somehow hover just above the event horizon.

In chapter 6, we met S5 0014+813, a candidate for the most massive black hole known, with a mass equivalent to 40 billion Suns. The gravitational pull of this gargantuan object should far exceed any white dwarf or neutron star, right?

But surprisingly, the acceleration due to gravity above the event horizon of this black hole is a mere 840 miles per hour per second. Sure, this is more than 100 times the gravity of the

Earth, but it's more than a billion times weaker than the gravity of a neutron star. How can a mighty black hole have such puny gravity? What's going on?

The answer lies in the fact that black holes are on a continual growth spurt.

Take some dough, and roll it into a ball 4 inches across. Now make another ball, exactly the same size. If you now squash the two balls together and make a single larger ball, the new ball will be double the mass and double the size of the two original balls. No surprises there, but note that "double the size" means double the volume, not double the diameter. The new ball will be twice as large, but its diameter will be 5 inches, only 26% more than the two smaller balls.

But black holes are strange objects, and behave very differently from dough balls. If two black holes of the same mass collide and merge, the new black hole will have twice the mass *and* twice the diameter of either of the original black holes. As a black hole sucks in material and grows in mass and size, it expands in girth much faster than you'd expect. And because the strength of gravity of an object increases with mass but decreases as the square of its diameter, a black hole's gravity actually gets weaker as the hole gets bigger. S5 0014+813 might be extraordinarily massive, but it is also unexpectedly large—the diameter of its event horizon is more than 120 billion miles. It might be counterintuitive, but it's the black holes with the lowest masses and the smallest sizes that we expect to have the strongest gravity.

The lightest black holes we know of are "stellar black holes," the puny cousins of the supermassive black holes we encountered in chapters 5 and 6. In chapter 3 we discussed how massive stars explode as supernovas, leaving behind spinning neutron stars at their cores. But occasionally, when an especially large star explodes, the core is sufficiently massive that it can collapse to form a stellar black hole. If this happens to an isolated star, the resultant black hole will be invisible and undetectable from our vantage point here on Earth. But if the star is one half of a binary system, then after the supernova we are left with a normal star orbiting a stellar black hole. The black hole begins to tear streams of gas off its companion with its gravity, and the spiraling disc of heated gas glows brightly as it swirls down the hole's throat.

There are almost 20 cases in the Milky Way for which we can see this bright emission with our telescopes. Through detailed analysis, scientists can determine the properties of these black holes and their harried companions. It's this kind of analysis that has helped astronomers pinpoint the stellar black hole with the smallest known mass, and therefore the strongest gravity. The smallest black hole we know of is located some 8,000 light-years away in the constellation of Perseus, and is called "GRO J0422+32."

GRO J0422+32 was discovered by American astronomer Bill Paciesas on August 5, 1992, when it suddenly increased in brightness. Years of detailed study ensued, from which astronomers have determined that this black hole has a mass just under four times that of the Sun—about 9,000 trillion trillion tons. Based on this mass, we can calculate that GRO J0422+32 is 14 miles across, about the same size as a neutron star. But because GRO J0422+32 has a mass almost three times more than a neutron star, its gravity is correspondingly three times more powerful.

Stronger than anywhere else we know of in the Universe, the gravitational pull just above the event horizon of GRO J0422+32 is around 9 trillion miles per hour per second. If you somehow managed to drop a stone into this black hole from a vantage point of 30 feet above the event horizon, it would take just two-millionths of a second to fall in, and would pass the horizon at a speed of 20 million miles per hour.

GRO J0422+32 is the smallest black hole we know of, but are there smaller black holes out there, waiting to be discovered? Probably. Scientists have calculated that the lightest a stellar black hole can possibly be is around three times the mass of the Sun. There are likely to be such objects out there, but we may have to look hard to find them. What's more, if we look beyond black holes formed in supernova explosions, there is no practical lower limit to the size or mass such an object can have. It has even been suggested that many tiny black holes, each with a mass of a mere billion tons, could have formed soon after the Big Bang.

No evidence for such featherweight black holes has yet been found, and if they do exist, they would be smaller than an atom. But if there are any of these black holes out there, their gravitational pull would be beyond spectacular.

The long slow dance

Black holes represent one extreme of the gravity spectrum, so what lies at the other end? How weak can gravity get?

In one sense this is an impossible question to answer. Since every object in the Universe exerts gravity on every other object, the gravitational pull you experience from a single atom at a distance of 10 billion light-years is unimaginably weak, and is completely swamped by all the much nearer things (the Earth, the Sun, the Milky Way) that continually tug on us.

What's more, while gravity is only ever an attractive force and never a repulsion, if you are halfway between two objects of identical mass, then their pulls due to gravity are equal but opposite, and the total gravitational force that you then experience is zero. Therefore, there are likely to be many places throughout the cosmos where the net gravity is negligible, because the attractions of different objects in different directions cancel each other out. Because of this, it is not especially meaningful to ask where in the Universe gravity is the weakest.

But what if we rephrase the question, and only consider situations in which a single object clearly exerts its gravitational pull on another object, without being swamped by all the other sources of gravity elsewhere in the cosmos? For example, the Earth feels the gravity of all the stars in the Milky Way, and of other galaxies, and of distant galaxy clusters. But these are dominated by the gravity of the Sun, and so as a result the Earth's dominant motion is to orbit our nearest star.

As we consider cases of weaker and weaker gravity, such orbits become increasingly difficult to maintain. The hold of the central object becomes ever more fragile in competition with all the other gravitational pulls in the Universe, until it loses the ability to keep surrounding bodies on their paths.

So let's ask this question: What is the gentlest pull that any object in the Universe exerts, and yet still is able to force another body to orbit it?

Such situations are likely only to arise in relatively quiet, uncluttered regions of the cosmos, where orbits are long, languid, and undisturbed. In more complicated regions, any weak gravitational bond between two objects will quickly be interfered with and disrupted by the stronger gravity of an interloping third party. For example, in the frenzied environment of a globular cluster (see chapter 2), any two stars with weak gravity will never be able to establish orbits around each other, since there will always be hundreds of other stars in the neighborhood, all with their own gravitational agendas.

So in looking for the weakest gravity, we should start by hunting for large, slow, isolated orbits.

An obvious initial candidate is the orbit of the Earth around the Sun. Our planet has shuttled around its central star more than 4 billion times without much interference. The gravitational acceleration with which the Sun holds the Earth in its orbit is indeed surprisingly weak: about 0.01 miles per hour per second, or a factor of 1,650 times smaller than the gravity the Earth exerts on its surface. This means that if you were to hover at the Earth's position and drop a stone down onto the Sun, it would take a full minute for the stone to fall its first 30 feet.

Just as the Earth orbits the Sun, the Sun orbits the center of the Milky Way. This galactic orbit is far longer, taking more than 200 million years for just one circuit. The gravity that the Milky Way exerts on the Sun is correspondingly far weaker, with a strength of less than one-billionth of a mile per hour per second. The Milky Way is immensely massive, but most of this mass is tens of thousands of light-years away. Over these large distances, the pull of gravity becomes exceedingly tenuous.

But 200 million years is still a relatively brisk orbit on cosmic scales: The oldest stars in the Milky Way have been able to complete such orbits more than 50 times since they began their lives. To find even longer orbits, under the influence of even weaker gravitational pulls, we need to go beyond the paths of individual stars in the Milky Way, and consider the orbits of entire galaxies.

There are numerous small galaxies in our neighborhood, many of which are in the gravitational thrall of our Milky Way. Some of these small galaxies are more than 500,000 light-years away from the Milky Way, and take around a billion years to complete an orbit. The gravity that the Milky Way brings to bear on these galaxies is almost 100 times weaker than the pull it exerts on our Sun.

But it is the Milky Way's most ambitious, most fragile gravitational entreaty that will eventually be its undoing.

On a dark fall night in the Northern Hemisphere, one can usually make out a gray elongated patch in the sky, a few times larger than the full Moon. This is the Andromeda Galaxy (also known as "Messier 31"), a giant spiral galaxy much like our own Milky Way, but at a distance of 2.5 million light-years. Andromeda and the Milky Way are the two biggest members of a collection of a few dozen galaxies known as the "Local Group." Both Andromeda and the Milky Way have their own retinues of small galaxies that orbit them. But incredibly, they are also both in a giant, protracted orbit around each other.

At this large separation, the gravity that holds together the Milky Way and Andromeda is extremely weak, with a corresponding gravitational acceleration of just 0.0000000000008 miles per hour per second. A stone dropped in this minuscule gravitational field, falling for 6 hours, would travel about the width of a human hair.

But the gravity that the Milky Way exerts on Andromeda, puny as it might be, will have dramatic consequences. Andromeda and the Milky Way are yet to complete even a single orbit around each other over the entire life of the Universe. Astronomers have estimated that a full loop would take around 17 billion years, but it is unlikely that this will ever happen. This is because the path followed by these galaxies in their orbit is not circular like the Earth's track around the Sun or the Sun's around the Milky Way, but has the shape of a highly elongated ellipse. In an elliptical orbit, there are points where the objects are widely separated, but others where they pass very close together. Right now, the Milky Way and Andromeda are midway between these two extremes, but time is bringing them ever nearer to the closest point of passage.

At the moment the Milky Way and Andromeda are plunging toward each other at a combined speed of 267,000 miles per hour, and are gradually picking up speed under the influence of the gentle gravitational acceleration between them. A billion years from now, the distance between the two galaxies will have halved, and Andromeda will loom as a spectacular sight in the night sky, four times brighter and four times larger than we see it today.

As time progresses, Andromeda will swell to fill almost half the sky. Will there eventually be a catastrophic collision as the two galaxies hit head-on? Not quite. Although a galaxy appears to be a swarming dense mass, the spacing between individual stars is enormous. So even if Andromeda and the Milky Way

plow into each other, there are unlikely to be any actual collisions between stars. It is more apt to imagine two jars of sand
being mixed together rather than to think of two cars colliding.

Furthermore, current calculations suggest that Andromeda
and the Milky Way are not exactly on a head-on trajectory, but
rather will sideswipe as they run past each other. This near
miss, about 2 billion years from now, will still bring enough
gravitational attraction to bear to tear entire spiral arms off both
galaxies, flinging stars outward in spectacular arcs 100,000
light-years long.

After this, the Milky Way and Andromeda will inexorably
continue in their orbits around each other, but now greatly
drained of energy and momentum by their initial passage.
Their next circuit around each other will be much smaller, and
they will lurch back together for a final embrace. Eventually all
the beautiful spiral arms will be lost, and the remnants of the
two galaxies will merge into a single giant ball of stars, forming
an elliptical galaxy. (Although the details of how all this will
play out are not certain, astronomers have sufficient confidence
in their prediction of the Milky Way's eventual fate that the
new hybrid galaxy has even been given a name: "Milkomeda.")

Any descendants of ours will have a spectacular view of
these events. The Sun's orbit will probably be shifted into a
wild, swinging trajectory: Over the course of millions of years,
this is likely to take us deep into the blindingly bright core of
the Milky Way, then out to a vantage point deep in space
where we can see the entire collision laid out before us on the
sky, and then back into the galactic core once more.

It is remarkable that so much chaos and disruption will
result from the gentlest of gravitational pulls between these

two galaxies, at a level 26 trillion times weaker than the gravity with which the Earth holds us to its surface. But the gravitational attraction between the Milky Way and Andromeda has billions of years ahead of it with which to do its work. Over this long span, even this incredibly weak gravity can have spectacular effects.

Barely hanging on

The Milky Way and the Andromeda Galaxy just lightly touch each other. But can we find an orbit in which the gravity is even weaker?

There are a huge number of waltzing pairs of galaxies throughout the Universe, just like the Milky Way and Andromeda. But our Galaxy and its neighbor are reasonably massive. All over the sky there are many smaller galaxies whose gravity is correspondingly weak. And if two low-mass galaxies can somehow come together in an isolated region of space, such that they can move without being affected by large galaxies like our own, they can then reach out with their feeble gravity and take up a fragile orbit around each other.

Of the many binary pairs of small galaxies in our catalog, the two known galaxies that are bound together most weakly are an obscure duo known as "SDSS J113342.7+482004.9" and "SDSS J113403.9+482837.4," or as we'll call them, Napoleon and Josephine. These two galaxies are 139 million light-years from Earth, high in the northern sky in the constellation of Ursa Major. They are 40,000 times too faint to see with the naked eye and, even through a telescope, they make an unimpressive couple. Each galaxy is about a thousand times less massive than the Milky Way, and both appear as unremarkable smudges on our deepest astronomical images.

However, what is surprising about these two galaxies is the weakness of the gravity with which they hold each other together in their orbit. The larger of the two, Napoleon, reaches across 370,000 light-years to its companion with a gravitational attraction of just 0.00000000000002 miles per hour per second, or about 900 trillion times smaller than an apple experiences when it falls from a tree. If you hovered at the position of Napoleon and dropped a stone toward Josephine, you would have to watch the stone for 50,000 years for it to accelerate up to a speed of about one inch per second, slightly faster than a garden snail. Wait another 4 million years or so, and it would move up to around walking speed.

Not surprisingly, with this incredibly weak gravity between them, these galaxies take an eternity to orbit around each other. In fact, in the billions of years since these two galaxies formed, they have probably passed through barely one-fifth of their first orbit.

And it's unlikely they will ever complete that orbit. The gravitational attraction between Napoleon and Josephine is so weak that it's merely a matter of time before some wandering interloper passes through their neighborhood, and uses its stronger gravity either to capture these two into its own orbit or to scatter this delicate pairing to the winds.

10

EXTREMES
OF DENSITY

If you've ever gone tenpin bowling, you'll know that the alley usually provides hundreds of bowling balls for you to choose from. The balls have a range of different masses, catering to everyone from small children to burly adults. But all the bowling balls sitting on the rack are the same size—the heavy balls and the light balls look essentially the same. And so before you begin your game, you need to hold a few balls in your hands, to try and find the one that suits you best.

But what is it about a heavy ball that makes it different from a light one, given that the two balls are the same shape and the same size? The difference is, of course, its density: The heavy ball packs more mass into the same volume, and so is denser.

We can easily put numbers on this. We can measure mass in ounces—1 ounce is about the mass of a slice of bread. And we can measure volume in cubic centimeters—in cooking, 1 teaspoon is about 5 cubic centimeters. Putting these together, density is mass divided by volume, and so is measured in ounces per cubic centimeter. For example, a 10-pound bowling ball has a density of 0.03 ounce per cubic centimeter, while a 16-pound bowling ball has a density of 0.05 ounce per cubic centimeter.

Bowling balls might feel heavy when dangling from three fingers, but if we look at the world around us, we can quickly establish that they are not especially dense. For example, iron has a density of 0.28 ounce per cubic centimeter, while gold has a density of 0.68 ounce per cubic centimeter. The naturally occurring substance with the highest density is osmium, which has a density of 0.8 ounce per cubic centimeter. Osmium is so dense that a bowling ball made of pure osmium would weigh more than 270 pounds (and would cost well over a million dollars!).

At the other extreme, at sea level the air we breathe has a density of just 0.00004 ounces per cubic centimeter, dropping to a density of around 0.00002 ounce per cubic centimeter at the top of Mount Everest. At sea level, even 10 gallons of air weighs little more than a sugar cube.

But beyond the bounds of Earth, what can the Universe offer with regard to density? As you would expect, the lowest and highest densities in the Universe are both far beyond our comprehension.

Crystalline spheres

The Earth itself is reasonably dense. The average density of our home planet is 0.2 ounce per cubic centimeter, ranging from around 0.1 ounce per cubic centimeter at the surface (comparable to the density of aluminum), up to 0.5 ounce per cubic centimeter at the Earth's core (denser than lead).

In comparison, the Sun is far more massive than the Earth,

but is also much larger. Combining these two effects, we find that the Sun is not especially dense, with an average density of about 0.05 ounce per cubic centimeter, only a little higher than the density of water.

The core of the Sun, where nuclear reactions take place under conditions of extreme temperature and pressure, is much denser than this average, around 5 ounces per cubic centimeter. At this density, a parcel of gas the size of a small pumpkin would weigh more than a ton.

There is clearly a large contrast between the watery consistency of most of the Sun and its compact core. However, in other stars the disparity is even more severe. In chapter 4, we discussed red giants, the elderly stars that have an extremely dense core surrounded by a huge, bloated outer envelope. Most of the gas in a red giant is at an extremely low density of around 30 billionths of an ounce per cubic centimeter. This is a thousand times less dense than the Earth's atmosphere: If you were to sit even deep inside a red giant star, you would conclude that you were surrounded by pure vacuum unless you first made careful measurements. So tenuous are red giant stars that it is difficult to define the concept of an actual surface that marks the interface where the star ends and surrounding space begins. Rather, the star slowly fades away into nothingness, analogous to driving out of a haze and gradually into clear skies.

In contrast, gravity ensures that the massive hot core of a red giant is far more compressed than any part of the Earth or Sun, with an estimated density of 220 pounds per cubic centimeter. To achieve this density, one would need to take an entire car and crush it down to the size of a golf ball.

When red giants finally exhaust their fuel and end their lives, they release their outer layers of gas into space, while their cores remain as hot, glowing embers. These central remnants are white dwarfs, stars whose extremes of temperature, magnetism, and gravity we have encountered in earlier chapters.

A white dwarf's enormous gravitational attraction results from the fact that such a star is the size of the Earth, but with 300,000 times the mass. These extraordinary properties also imply a huge density, of around 5,000 pounds, or 2.5 tons, per cubic centimeter. To visualize this density, imagine a giant set of scales, on one side of which we put 100 people. To balance the scales using white dwarf matter, we would need to place just a teaspoon's worth of material on the other side.

I noted above that osmium, with a density of a mere 0.8 ounce per cubic centimeter, is the densest of the naturally occurring elements. A white dwarf is almost 100,000 times denser than osmium, so just what are these objects made of?

Before a star becomes a white dwarf, it's made of ordinary gas: mainly free-floating hydrogen atoms, with smaller amounts of helium and other impurities mixed in. But as gravity squeezes the dying star's core, a strange transformation occurs. The star now becomes so dense that individual atoms can no longer move at all, and become held in place like orbs hanging from a Christmas tree. A white dwarf is therefore not a foaming, fiery ball of gas like other stars, but rather a solid sphere of material.

What's more, the positioning of the atoms in a white dwarf becomes extremely precise, with atoms placed in a regular, three-dimensional grid. We see this orderly arrangement of atoms under our microscopes all the time: Such a pattern is known as a crystal. Ordinary table salt is such a crystal, as is diamond. And so under the immense forces that gravity has imparted, a white dwarf takes the form of an enormous, super-dense, single crystal.

You might imagine such a crystalline star as a delicate, fragile structure. Given the ferocious pull of gravity, which relentlessly tries to squeeze a white dwarf to an even smaller size, what stops the crystal from shattering, and then collapsing to even higher densities?

To understand what holds a white dwarf in place, and prevents it from collapsing in on itself, we need to dig deep into our understanding of modern physics. For as we move to increasingly higher densities, the hidden world of quantum physics rears its head.

Quantum mechanics, a branch of physics first developed in the early years of the 20th century, describes the behavior of small particles such as protons and electrons. Its complex predictions are often bizarre—sometimes requiring particles to be in two places at once, or claiming that electrons can tunnel through solid barriers, magically emerging on the other side. But despite the strangeness of these calculations, as far as we can tell this is actually how the microscopic world works. So far, quantum theory has done a beautiful and near-flawless job of explaining how individual atoms behave.

Normally quantum mechanics has little bearing on large objects like people, planets, or stars. However, at the extreme densities of a white dwarf, some of the arcane aspects of quantum mechanics come into play. In particular, all matter in the Universe must heed a seemingly unbreakable rule known as the "Pauli Exclusion Principle." This rule, developed by the Nobel Prize–winning physicist Wolfgang Pauli, essentially forbids two electrons with the same energy from occupying the same place in space.

Every electron in every atom in your body fastidiously obeys this rule. However, it is normally an easy rule to comply with, because there is lots of space in which electrons can roam—the Pauli Exclusion Principle almost never comes into play. Inside a white dwarf things are different: The densities are so high that one comes up against a fundamental limit as to how many electrons one can pack into a given volume. The situation is analogous to a busy multistory parking garage: Eventually every spot on the first level is taken, and it becomes impossible to fit any more cars in. Cars that want to find a spot need to try the second level. Eventually, the second level fills up too, and the only parking spots left are on the third level, and so on.

In the same way, a white dwarf is so dense that we quickly establish a situation in which at the very core of the star there are electrons with all possible energies. Other electrons feel the pull of gravity trying to drag them down to the center of the star, but they are forbidden by the Pauli Exclusion Principle from occupying this space. These electrons cluster near the core, until all available spots allowed by Pauli's principle are taken in this region too. Other electrons fill up all the available slots a little farther out, and so on and so on, all the way out to the white dwarf's surface.

In this way all the electrons in a white dwarf are held in place, caught in the crossfire between gravity's unrelenting demand for them to fall to the star's center, and the Pauli Exclusion Principle, which forbids them from being packed any more tightly.

And so, at a density of 2.5 tons per cubic centimeter, and having pushed a fundamental law of physics to its limit, are white dwarfs the densest possible objects in the Universe? Amazingly, no. Even the density of a white dwarf is relatively tame by the extreme standards of the cosmos.

Great balls of pasta

Throughout this book, one group of celestial objects has perhaps made more appearances than any other. Neutron stars seem to epitomize the Universe's extremes: They spin incredibly rapidly (chapter 3), fly through space at extraordinary speeds (chapter 5), are the strongest magnets in the Universe (chapter 8), and have unbelievably strong gravity (chapter 9). And here they make a final appearance, for neutron stars also represent one of the densest states of matter known in the Universe.

To quickly recap, we learned in chapter 1 that massive stars are so hot that nuclear fusion eventually converts all the atoms in the stellar core into pure iron. However, this iron will not then fuse into heavier elements. Meanwhile, a hot shell of gas around this core continues to burn, adding more iron.

The doomed star's iron core therefore grows steadily larger and heavier as time passes. When this giant ball of iron hits a mass of 3,000 trillion trillion tons (about 1.4 times the mass of the Sun), it suddenly and catastrophically collapses into a neutron star, now only about 15 miles across. Until this moment, the iron core was holding up the rest of the star. The star's outer layers, their supporting platform now gone, fall inward until they slam into the rock-hard surface of the neutron star. The dramatic rebound off this surface drives a shock wave through the rest of the star, ripping it apart in an enormous supernova explosion.

But let's now have a closer look at the neutron star, this strange beast that represents both the final stage of the star's life and the trigger for its explosive ending. With a mass of 3,000 trillion trillion tons but a diameter of just 15 miles, the average density of a neutron star is around 375 million tons per cubic centimeter, or 150 million times more dense than a white dwarf! At this density, a fragment of a neutron star the size of a grain of sand would weigh more than a modern aircraft carrier. And a piece of neutron star the size of a die would weigh three times more than the entire human population.

How is such an unimaginably high density created and then sustained? What happened to the inviolable Pauli Exclusion Principle, which limits the density of a white dwarf to a paltry 2.5 tons per cubic centimeter? Do neutron stars not adhere to the laws of physics?

Indeed, neutron stars play by the same rules as everyone else. However, their extreme mass and gravity force them down a strange and unexpected path.

I explained earlier that the structure of a white dwarf is such that electrons simply cannot be packed together more tightly. My analogy was that of a multistory parking garage: Once every spot is taken, it is impossible to fit any additional cars into the building. However, this is a fundamental limit only if we restrict ourselves to an orderly approach, in which cars park in the marked spaces, with a gap between each car and the ones on either side.

Suppose that we abandon all caution, and park the cars so close together that they scrape up against each other; indeed, so close that no passenger could ever open the door and get in or out. After we've squeezed in more cars this way, let's go further, and fill up all the lanes and ramps with more cars. If the ceilings of each level are high enough, then we can fit in yet more cars by piling them on top of each other.

We now have a garage that holds substantially more cars than any normal definition of a full lot. But let's keep going. Even if we can't possibly fit any more cars in, if we look carefully there's still a lot of unoccupied space. Each car itself contains a lot of empty space, enough room for four or five passengers, plus some groceries and shopping in the trunk. If we crush each car flat, then we can use this space too, and fit even more cars in. We can keep pushing cars into the lot until the entire building is packed solid with steel, rubber, and plastic, with no air or empty space remaining. You might imagine that in this way we can fit perhaps 10 times as many cars into a parking garage as it could hold if the cars were parked in normal fashion in their allocated spaces.

In the same way, a neutron star achieves far higher densities than a white dwarf, because it abandons the usual rules of how atoms and electrons are ordered.

In the normal matter of which you and I are composed, each atom consists of a tiny nucleus of protons and neutrons, surrounded much farther out by a cloud of orbiting electrons. But between the nucleus and the electrons of an atom is a huge, unoccupied gap. A seemingly solid object, composed of an enormous number of atoms, is on closer examination merely the occasional nucleus, each surrounded by a far-off cloud of electrons, with nothing in between. So while you might feel substantial, it turns out that approximately 99.9999999999997% of your body is empty space!

In the same way, a white dwarf represents the most tightly packed possible form of normal matter, but even at a density of 2.5 tons per cubic centimeter, it still contains a lot of unused space. Neutron stars are the only objects in the Universe, apart from atomic nuclei themselves, which fill in these spaces, and they do it in a surprising way.

The iron core of a massive, dying star finds itself in a similar situation to a white dwarf. Every atom is pushed up against each other, their individual clouds of electrons overlapping, so much so that the Pauli Exclusion Principle comes into play. Two electrons are forbidden from being in the same place with the same energy, and this limit temporarily halts the inexorable inward pull of gravity, preventing the core from collapsing in on itself. But so massive is this iron core that gravity eventually finds a way to proceed. In a catastrophic process known as "neutronization," many of the protons and electrons in the star merge, forming neutrons. With 90–95% of the electrons now gone, the empty spaces that they were protecting around each atom's nucleus become available, and the star can dramatically shrink in volume. Rather than being enormously separated as in normal matter, the newly formed neutrons can move together so closely that they essentially touch. The density of the star skyrockets, until we reach an incredible 375 million tons per cubic centimeter.

What's to stop this neutron star from collapsing further under gravity? The final barrier against the ultimate collapse is once more the Pauli Exclusion Principle, but this time applied to neutrons rather than electrons. No two neutrons can simultaneously have the same energy and the same position. Forced to obey this universal law, the neutrons are held against gravity in a tightly packed grid, under conditions of almost unimaginable pressure and density.

While white dwarfs are believed to have a crystalline structure, astronomers' calculations suggest that neutron stars are markedly different beasts.

Just like the Earth, a neutron star has a gaseous atmosphere and a solid surface. However, the similarities end there.

First, the intense gravity of a neutron star (see chapter 9) ensures that its atmosphere is only an inch or two thick. The surface of the neutron star is home to many normal atoms, atoms which escaped the neutronization that encompassed most of the star at its formation. As a consequence, this material is not especially tightly packed, with a modest density of just 0.2–0.4 ounce per cubic centimeter, about the same density as iron. Generally we think of normal atoms as spheres, analogous to tiny marbles or billiard balls. However, the intense magnetism at the surface of a neutron star (see chapter 8) squeezes individual atoms into elongated cylinders, making them into tiny compasses in their own right. As a consequence, like a chain of paper clips dangling from a toy magnet, adjacent atoms can join up end to end. The result is that atoms on a neutron star's surface organize themselves into spectacularly long chains, beautifully aligned with the direction of the magnetic field.

Because all the atoms on the neutron star's surface are now interlinked and well ordered, this bizarre material is unbelievably strong, and is virtually impossible to snap or break. In comparison, consider a plastic ruler. This plastic is remarkably tough: If you try to cut the ruler in half with a pair of scissors, you may not be able to do so. However, if you bend the ruler, eventually it will snap in two.

The plastic in a ruler, like the atoms on a neutron star's surface, is composed of long chains of atoms. These chains are individually extremely strong, but there are many gaps and imperfections between them. When you bend the ruler, it's along these cracks that the plastic eventually splits apart. However, the atoms of a neutron star's crust are so tightly and rigidly organized along their chains that fracturing the crust requires not the separating of badly joined chains, but the breaking of individual chains themselves, something that requires far more energy and effort. Nuclear physicists Charles Horowitz and Kai Kadau have carried out detailed calculations on the strength of a neutron star's crust, and have found that this material is 10 billion times stronger than steel!

The crust of a neutron star is about half a mile thick. Below this layer, we think things get even stranger, as we move into the bizarre world of "nuclear pasta." Normally protons and neutrons are held together tightly in an atomic nucleus by an intense attraction known as the "strong nuclear force." These nuclei can be thought of as hard spheres, like tiny ball bearings. However, there's potentially another force at play, stemming from the fact that a proton has a positive electrical charge. Two positive charges repel each other, so two protons brought near each other will push back and try to separate.

Inside an atomic nucleus, the strong nuclear force is (as its name suggests) so strong that it easily overcomes the repulsion that two or more protons in that nucleus might exert on each other. Despite their distaste for one another, the protons are forced to sit side by side.

But two protons from two separate nuclei are not held together by the strong nuclear force. They will do whatever they can to stay as far away from each other as possible. Normally this repulsion never comes into play, because two nuclei are so widely separated, each with a vast empty void around them, with electrons circling on the outside. Inside a neutron star, however, most of these electrons have been neutronized, and individual nuclei are pushed far closer together than they would like. Individual nuclei now almost touch, and protons from different nuclei try to fiercely repel each other. But there is nowhere to move, and this repulsion instead puts the nuclei under enormous strain, squashing them into strange shapes.

As I noted above, the simplest way to envision the shape of the nucleus in a normal atom is as a ball bearing. However, as one descends into a neutron star's depth, individual nuclei squeeze themselves into elongated tubes. At even higher densities farther into the star's interior, the nuclei are squashed into broad, flat sheets. American astronomer David Ravenhall was one of the first to appreciate this behavior, and in 1983 he suggested a set of gastronomical terms to describe this, now known as the "pasta sequence." Normal atomic nuclei are spheres, like meatballs. As they stretch themselves into long tubes, they become spaghetti. And as they flatten themselves into sheets, they are transformed into lasagna!

As one progresses to even higher densities, the sequence continues. The individual nuclei now begin to merge, with only narrow hollow tubes separating them: nuclear ziti, of course. At yet more extreme densities, only small spherical holes persist, making ravioli. And finally, as one moves into the densest parts of the neutron star, all the nuclei begin to overlap and merge into a single giant, continuous, atomic nucleus. The inevitable conclusion is that the central regions of a neutron star, with a density of hundreds of millions of tons per cubic centimeter, contain the sauce.

Lighter than air

Neutron stars are extraordinary objects, but they are not the final state into which a dying star can collapse. Usually when a massive star explodes, the core survives as a neutron star, but as we saw in chapter 9, a very large star can leave behind a heavier core after its supernova explosion, which collapses further into a stellar black hole. Black holes have gravity so extreme that from inside their event horizon, not even light can escape.

Since some stellar black holes are even heavier and smaller than neutron stars, you'd expect their interiors to reach even higher densities. This is indeed the case, but this first needs some further thought and explanation.

I need to caution that it isn't entirely clear what it even means to talk about the density of a black hole. Inside a black hole's event horizon, we don't expect material to be composed of normal atoms like you or me, or even exotic nuclear pasta as thought to reside inside neutron stars. While we can't be completely certain, consensus is that all the mass of a black hole is compressed into a "singularity," an infinitely small mathematical point at the very center of the event horizon. If so, then all black holes technically have infinite density.

Unfortunately, this isn't easy to comprehend, nor do we yet have the measurements to test this idea. Instead, we usually describe a black hole in terms of its "equivalent density"—that is, we use the size of the black hole's event horizon to determine its volume, and then divide mass by volume to calculate its density. This isn't a real density, in the sense that you can't scoop out a spoonful of black hole material and place it on a scale to weigh it. However, if you think about taking some ordinary matter and squeezing it until it is small enough to become a black hole, the density to which you have to compress material to create a black hole is the equivalent density.

A second strange aspect of black hole densities follows from our discussion of a black hole's gravity in chapter 9. In the same way that increasingly more massive black holes have progressively weaker gravity, heavier black holes also have a much lower density than light ones.

So in our quest to find the black hole with the highest possible density, we need to look to the black hole with the lowest mass. Just as in chapter 9, the current title holder is GRO J0422+32, the lightest known black hole, with a mass just under four times the mass of the Sun (about 9,000 trillion trillion tons). The event horizon of GRO J0422+32 has a diameter of about 14 miles. So dividing mass by volume, we can calculate that the equivalent density of GRO J0422+32 is a staggering 1.3 billion tons per cubic centimeter, more than triple the density of a neutron star.

Only a lightweight black hole like GRO J0422+32 can reach such a high density. There are about a dozen stellar black holes known with masses around 8–12 times that of the Sun. Even this modest increase in mass causes a substantial decrease in equivalent density: A black hole weighing 10 times the mass of the Sun has a density of 200 million tons per cubic centimeter, one-sixth of that for GRO J0422+32, and about half the density of a neutron star.

As we have seen in chapters 5 and 6, there are many black holes that are far larger and heavier than those left behind in supernova explosions. At the centers of galaxies we often find supermassive black holes, weighing millions or billions of times the mass of the Sun. Despite their enormous mass, supermassive black holes have such large event horizons that their equivalent densities are surprisingly low.

Let's start with our own local supermassive black hole, Sagittarius A* at the center of the Milky Way. As we saw in chapter 6, astronomers have made precision measurements of the mass of Sagittarius A*, using the behavior of the stars we can see orbiting close in around it. Our current best estimate from these studies is that Sagittarius A* weighs 4.3 million times the mass of the Sun.

Astronomers haven't yet been able to definitively measure the diameter of Sagittarius A*, but for a mass of 4.3 million Suns, we expect the event horizon to be 15 million miles across, smaller than the orbit of Mercury around the Sun. Since we know the mass of Sagittarius A* and we know its expected size, we can calculate its density, and the answer is surprisingly small: The density of this supermassive black hole is just 2.3 pounds per cubic centimeter. This is impressive by everyday standards, but not by astronomical ones: It is 300 billion times less dense than a neutron star, 0.05% of the density of a white dwarf, and only six or seven times the density of gas at the Sun's core.

We saw in chapter 6 that there are many supermassive black holes far more massive than Sagittarius A*. A more typical supermassive black hole has a mass of 100 million Suns, and the equivalent density is then only 0.04 ounce per cubic centimeter, about the same as that of water! So while a stellar black hole is thought to be produced by a severe collapse and compression of a dying star, no such stressful processes are needed to make a supermassive black hole. If you had a large enough bathtub, you could simply pour water into it until the mass of the water hit 100 million times the mass of the Sun, and you would now have enough mass in a small enough volume to make a black hole. As German astronomer Heino Falcke warns his students, don't leave the faucet running when you go on a long vacation, lest catastrophe strike!

And what about the largest black holes known? In chapter 6, I explained that the most massive black hole yet discovered is S5 0014+813, with a mass of 40 billion Suns. Although this mass is impressive, the equivalent density of this monstrous beast is a tiny three-millionths of an ounce per cubic centimeter, roughly the same density as that for gas inside a fluorescent light tube.

There is a temptation to imagine the largest black holes as places of extraordinarily extreme densities. Indeed they are, but in the exact opposite sense from what you might expect. S5 0014+813 is nothing more than a gargantuan cosmic helium balloon, far lighter than air.

Bubbles of nothing

The supermassive black hole S5 0014+813 might have a surprisingly low density, but the Universe can produce far more remarkable extremes at this small end of the scale.

Air is probably the lowest density material we typically encounter on a day-to-day basis, with a density of 0.00004 ounce per cubic centimeter. However, by cosmic standards, 0.00004 ounce per cubic centimeter is still an extremely high density.

To explore even lower densities, it is helpful to phrase things in a different way. Instead of measuring density in ounces per cubic centimeter, we need to talk about "number density," that is, the number of atoms contained in each cubic centimeter. By this measure, air is positively packed with matter, with a number density of more than 40 million trillion atoms per cubic centimeter.

As a comparison, let's consider the dark nebulas that we discussed in chapter 2, so choked with dust that they completely prevent any light from penetrating them. Despite their dense, dingy appearance, these dark nebulas have a number density of under a million atoms per cubic centimeter. This density is so low that a dark nebula comparable in size to an Olympic swimming pool would weigh less than a billionth of an ounce.

This is an extremely low density, but even here on Earth we can do better. For hundreds of years, scientists found clever ways to push to increasingly small number densities, creating ever more rarefied environments. The current state of the art, involving experiments that take several months to perform, results in a number density of only 500–1,000 atoms per cubic centimeter. By all reasonable measures, a gas in this state is a near-perfect vacuum.

But the Universe can deliver far lower densities than this. In the Milky Way, dark nebulas are the exception—most of the gas clouds scattered throughout our Galaxy have typical number densities of just 1 atom per cubic centimeter. This density is so low that if you were to replace all the water in all the world's oceans with gas of this density, the total mass of gas required would be about a tenth of an ounce!

Outside the Milky Way, in the empty reaches of intergalactic space, the density is even lower. In these isolated expanses between galaxies, the typical number density is just 0.00001 atom per cubic centimeter, meaning that individual atoms are now around a yard apart. A cloud of such gas the size of the entire planet Earth would weigh less than a thousandth of an ounce.

You might think that at this extremely low density all this gas is just a mere footnote in the cosmos's inventory of material. However, so large are the volumes between galaxies that the total mass of this intergalactic gas comfortably exceeds the combined weight of all the planets, stars, galaxies, and clusters in the Universe. The densities we are used to dealing with in everyday astronomy, let alone in everyday life, are the extreme exceptions rather than the norm.

This low density of intergalactic gas is hard to comprehend. But even this is well above what is found in the emptiest parts of the cosmos.

As we learned in chapter 4, galaxies are not scattered uniformly throughout the Universe, but are arranged into a spectacular filigree of sheets, filaments, shells, and cavities. The walls of these intergalactic soap bubbles are busy concentrations of stars and galaxies, filled with all the different forms of activity and energy that we have explored in preceding chapters. However, the interiors of these bubbles are unimaginably, frighteningly empty.

These "cosmic voids" do not contain galaxies. They are free from stray stars or planets. In these vast wastelands, often stretching across space for more than 100 million light-years, there is nothing more than the occasional lone atom of hydrogen.

So just how low does the density get in these emptiest of regions? The number density of a typical void is an incredibly tiny 0.00000002 atom per cubic centimeter. This is so sparse that even in a volume the size of a large room, you would be lucky to find a single atom. Or to put it another way, if you were to take one of the bowling balls we were considering at the start of this chapter, and grind it up into its individual constituent atoms, you would have to spread these atoms over a volume 4 million miles across to achieve the same density as is found in a cosmic void.

Such an environment seems extraordinarily alien, not just in comparison to our everyday earthly standards, but even to the emptiness of space that we find in and around the Milky Way. But the massive censuses of the Universe that have been undertaken over the last 10 or 20 years have now revealed the true story: These voids occupy around 90% of the volume of the Universe, with everything else in the margins.

If a visitor arrived in our Universe from some other cosmos, without any preconceived notions about what was important or interesting, they would quickly conclude that this was an uninteresting, bland Universe essentially filled with nothing but emptiness. Unless they were especially careful and observant, they might overlook some of the minor footnotes, the occasional regions of high density where stars can form, planets can orbit, life can emerge, great works of art can be painted, and majestic symphonies can be composed.

EPILOGUE

You might think that this brings us to the end of the story. But this is astronomy that we're talking about—there's not likely to ever be a final word to the tale of the cosmos.

Every mystery that astronomers are able to solve raises a dozen fresh questions that then need answers. Each new breakthrough or discovery quickly shows that we don't understand things quite as well as we thought. And every new telescope pointed at the sky reveals unexpected objects and phenomena that until then no one had been able to see.

The knowledge that there's so much more to do is both reassuring and exciting. Working as an astronomer, I often feel like a kid on a fantastic summer vacation, not wanting it to end. But while all vacations must eventually be followed by a return to the real world, the great cosmic adventure has no end in sight. By many measures we have learned an enormous amount about the Universe, but in other ways we have barely scratched the surface. The golden age of astronomy is only now just beginning.

So what is the future likely to hold? As we build bigger and more powerful telescopes, we will continue to find ever more stars, galaxies, clusters, and other celestial phenomena. Most of the objects that astronomers will discover will be similar to the ones we already know about, but a small handful will be record breakers in their own right, pushing past even the extremes I've described here. Somewhere out there are gas clouds even colder than the Boomerang Nebula, galaxies even bigger than IC 1101, musical notes even deeper than that which thrums through Abell 426, magnets even stronger than SGR 1806-20, and gravity even more powerful than for GRO J0422+32.

There are also so many amazing things that we already know about, but that I simply haven't had time or space to cover. There's the Leo Ring, an incredible loop of glowing gas 650,000 light-years across, which takes 4 billion years to orbit the galaxy it envelops. I could write another entire book just on dark matter and dark energy, two mysterious phenomena that together comprise 95% of all the mass and energy of the Universe. And even while I was writing *Extreme Cosmos*, a new record was set for the most distant object ever discovered, a galaxy known as "UDFj-39546284" at a distance of 13.2 billion light-years.

If there's a single unifying conclusion we can draw from the extraordinary objects we've found throughout the Universe (and from the knowledge that there's so much more still to be discovered), it's an appreciation of what a small and marginal role we play in the Universe's evolution.

Our entire Milky Way Galaxy is an unimportant, minute fleck on the celestial stage, an irrelevant region of bright light and dense gas, hidden amid the vast darkness and emptiness of the cosmic voids. On the other hand, what a triumph of pure thought it has been, that we mere humans, in the space of little more than a hundred years, have established how stars are born, live, and die; how galaxies evolve; and how the whole structure of galaxies, galaxy clusters, and cosmic voids fits together.

The cosmos is unquestionably extreme, and the numbers that measure these extremities can at first seem hard to comprehend. However, on closer inspection, the Universe's extremes become not only comprehensible, but turn out to be the vital keys needed to unlock the true wonder and elegance of the heavens.

Despite the seemingly hopeless mismatch between our limited human imaginations and the size and complexity of the Universe, it's astonishing that we think we understand so much of what we see. As baffled and cowed as we often find ourselves when confronted by the cosmos, it is perhaps humanity's ultimate accomplishment that we nevertheless can explain and appreciate the grandeur of the night sky.

EXTREME
EXPERIENCES

The tables that follow summarize the properties and records held by some of the objects and phenomena described throughout this book.

1. Extremes of Temperature

OBJECT	TEMPERATURE
Boomerang Nebula	−458°F
Cosmic microwave background	−454.76°F
Surface of the Sun	9900°F
White dwarf inside the Red Spider Nebula	540,000°F
Core of the Sun	27,000,000°F
Supernova explosion	9,000,000,000°F
Universe, 1 second after the Big Bang	18,000,000,000°F
Universe, 0.00000000001 second after the Big Bang	18,000,000,000,000,000°F
Universe, 0.0000000000000000000000000000000000000000001 second after the Big Bang	180,000,000,000,000,000,000,000,000,000,000°F

2. Extremes of Light

OBJECT	PROPERTY
Center of a dark nebula	Brightness: 1,000,000,000,000 times darker than the night sky as seen from Earth
Center of a globular cluster	Brightness: same brightness as the full Moon over the entire sky
Supernova explosion	Luminosity: 1,000,000,000 times more powerful than the Sun
Gamma-ray burst GRB 080319B	Luminosity: 100,000,000,000,000,000 times more powerful than the Sun

3. Extremes of Time

OBJECT	PROPERTY
Universe	Age: 13.7 billion years
Milky Way Galaxy	Age: approx. 13 billion years
Metal-poor star SDSS J102915+172927	Age: approx. 13 billion years
Sun, Earth, and solar system	Age: 4.6 billion years
Betelgeuse	Total life expectancy: 10 million years
Sun	Total life expectancy: 10 billion years
Red dwarf stars	Total life expectancy: 1 trillion years
Neutron star PSR J1748-2446ad	Rate of rotation: 716 times per second
Neutron star PSR J1909-3744	Roundest orbit known, perfect circle to within a thousandth of an inch

4. Extremes of Size

OBJECT	SIZE
Asteroid 2008 TS26	2–3 feet
Neutron star	15 miles
White dwarf stars	6,000–7,000 miles
Earth	7,926 miles
Red dwarf stars	120,000 miles
Sun	860,000 miles
Achernar	9 million miles
Betelgeuse	1 billion miles
WOH G64	1.2 billion miles
Milky Way Galaxy	100,000 light-years (600,000 trillion miles)
Elliptical galaxy IC 1101	5 million light-years
Sloan Great Wall	1.4 billion light-years

5. Extremes of Speed

OBJECT	SPEED
Earth (orbit around Sun)	66,000 mph
Mercury (orbit around Sun)	105,000 mph
Exoplanet WASP-12b (orbit around parent star 2MASS J06303279+2940202)	528,000 mph
Sun (orbit around Milky Way)	568,000 mph
Hypervelocity star SDSS J090745.0+024507	1,500,000 mph
Neutron star RX J0822.0-4300	3,500,000 mph
Oh-My-God Particle	670,616,629.13 mph (99.9999999999999999999996% of speed of light)
Speed of light	670,616,629.13 mph

6. Extremes of Mass

OBJECT	MASS
Mercury	360 million trillion tons
Earth	6.6 billion trillion tons
Jupiter	2 trillion trillion tons
Red dwarf GJ 1245C	7.4% of mass of Sun (160 trillion trillion tons)
61 Cygni B	63% of mass of Sun
61 Cygni A	70% of mass of Sun
Sun	2,200 trillion trillion tons
Sirius	2 times mass of Sun
Canopus	8 times mass of Sun
Alnilam	40 times mass of Sun
Wolf-Rayet star A1	116 times mass of Sun
Wolf-Rayet star WR 102ka	Initial mass 150–200 times mass of Sun
Population III star	300–500 times mass of Sun
Supermassive black hole Sagittarius A*	4.3 million times mass of Sun
Supermassive black hole S5 0014+813	40 billion times mass of Sun
Milky Way Galaxy	1 trillion times mass of Sun
Elliptical galaxy IC 1101	100 trillion times mass of Sun
Virgo galaxy cluster	1,000 trillion times mass of Sun
Galaxy cluster Abell 2163	4,000 trillion times mass of Sun
Observable Universe	approx. 400 billion trillion times mass of Sun

7. Extremes of Sound

OBJECT	PROPERTY
Supernova explosion	Volume: 330 decibels
Galaxy cluster Abell 426	Pitch: B flat, 56 octaves below middle C Volume: 170 decibels
Universe 10 years after Big Bang	Pitch: F sharp, 35 octaves below middle C Volume: 90 decibels
Universe 380,000 years after Big Bang	Pitch: C, 54 octaves below middle C Volume: 120 decibels

8. Extremes of Electricity and Magnetism

OBJECT	PROPERTY
Earth's surface	Magnetic field: 0.5 gauss
Sunspots	Magnetic field: approx. 1,000 gauss
Flare star YZ Canis Minoris	Magnetic field: 3,000–4,000 gauss
Babcock's Star	Magnetic field: 34,000 gauss
Tesla Hybrid Magnet, Florida	Magnetic field: 450,000 gauss
Multishot Magnet, New Mexico	Magnetic field: 889,000 gauss
MC-1, Russia	Magnetic field: 28 million gauss
White dwarf PG 1031+234	Magnetic field: 1 billion gauss
Neutron star PSR J1847-0130	Magnetic field: 100 trillion gauss
Magnetar SGR 1806-20	Magnetic field: 1,000 trillion gauss
Holifield Radioactive Ion Beam Facility, Tennessee	Voltage: 32 million volts
Neutron star PSR J0537-6910	Voltage: 38,000 trillion volts
Supermassive black hole	Voltage: 10 million trillion volts
Lightning bolt	Current: 20,000–50,000 amps
Earth's auroras	Current: approx. 1 million amps
Z Machine, New Mexico	Current: 26 million amps
Sunspots	Current: approx. 1 trillion amps
Neutron star PSR J0537-6910	Current: 1,000 trillion amps
Jets of radio galaxies	Current: approx. 1 million trillion amps

9. Extremes of Gravity

OBJECT	ACCELERATION DUE TO GRAVITY
Orbiting galaxies SDSS J113342.7+482004.9 and SDSS J113403.9+482837.4 (Napoleon and Josephine)	0.00000000000002 mph/second
Orbiting galaxies Milky Way and Andromeda	0.0000000000008 mph/second
Milky Way (at position of Sun)	0.0000000006 mph/second
Sun (at position of Earth)	0.01 mph/second
Moon (at surface)	3.6 mph/second
Earth (on board the International Space Station)	20 mph/second
Earth (at surface)	22 mph/second
Sun (at surface)	615 mph/second
Supermassive black hole S5 0014+813 (just above event horizon)	840 mph/second
White dwarf (at surface)	6.5 million mph/second
Neutron star (at surface)	3 trillion mph/second
Stellar black hole GRO J0422+32 (just above event horizon)	9 trillion mph/second

10. Extremes of Density

OBJECT	DENSITY
Cosmic voids	0.00000002 atom per cubic centimeter
Intergalactic gas	0.00001 atom per cubic centimeter
Interstellar gas	1 atom per cubic centimeter
Best laboratory vacuum	500–1,000 atoms per cubic centimeter
Dark nebulas	Up to 1 million atoms per cubic centimeter
Red giant (outer envelope)	0.00000003 ounce per cubic centimeter (600,000 trillion atoms per cubic centimeter)
Supermassive black hole S5 0014+813 (equivalent density)	0.000003 ounce per cubic centimeter
Air (at sea level)	0.00004 ounce per cubic centimeter
Sun (average)	0.05 ounce per cubic centimeter
Earth (surface)	0.1 ounce per cubic centimeter
Earth (average)	0.2 ounce per cubic centimeter
Iron	0.28 ounce per cubic centimeter
Earth (core)	0.5 ounce per cubic centimeter
Osmium	0.8 ounce per cubic centimeter
Sun (core)	5 ounces per cubic centimeter
Red giant (core)	220 pounds per cubic centimeter
White dwarf	2.5 tons per cubic centimeter
Neutron star	375 million tons per cubic centimeter
Stellar black hole GRO J0422+32 (equivalent density)	1.3 billion tons per cubic centimeter

FURTHER READING

For the interested reader, I have listed below some of the articles, books, and web pages that I used to assemble the information in *Extreme Cosmos*, or which I felt provided a helpful overview of a particular topic. Free prepublication versions of most articles in astronomy can be found at arXiv.org.

1. Extremes of Temperature

The hottest known white dwarf, inside the Red Spider Nebula: Pottasch, S & Bernard-Salas, J (2010) "Planetary nebulae abundances and stellar evolution II," *Astronomy & Astrophysics*, 517: 95.

The Boomerang Nebula: Sahai, R & Nyman, L (1997) "The Boomerang Nebula: The coldest region of the Universe?," *The Astrophysical Journal*, 487: L155.

2. Extremes of Light

I calculated the gradual disappearance of the stars in the sky as we are enveloped by a dark nebula using the results in the following two articles:

> » Alves, J et al. (2001) "Internal structure of a cold dark molecular cloud inferred from the extinction of background starlight," *Nature*, 409: 159.

» Román-Zúñiga, C et al. (2007) "The infrared extinction law at extreme depth in a dark cloud core," *The Astrophysical Journal,* 664: 357.

GRB 080319B, the naked-eye gamma-ray burst of March 19, 2008: Bloom, J et al. (2009) "Observations of the naked-eye GRB 080319B: Implications of nature's brightest explosion," *The Astrophysical Journal,* 691: 723.

3. Extremes of Time

SDSS J102915+172927, the lowest metallicity star: Caffau, A et al. (2011) "An extremely primitive star in the Galactic halo," *Nature,* 477: 67.

PSR J1748-2446ad, the fastest spinning star: Hessels, J et al. (2006) "A radio pulsar spinning at 716 Hz", *Science,* 311: 1901. A web search will provide stories and links to an object known as "XTE J1739-285," which in 2007 astronomers claimed was rotating 1,122 times per second, even faster than PSR J1748-2446ad (see www.esa.int/esaSC/SEMPADBE8YE_index_0.html). However, the data that led to this discovery have since been reanalyzed by other astronomers, who have been unable to verify the original result.

PSR J1909-3744, whose orbit is the most perfect circle known: Jacoby, B et al. (2005) "The mass of a millisecond pulsar," *The Astrophysical Journal,* 629: L113.

4. Extremes of Size

Mira and the trail of gas that it has left behind: Martin, D et al. (2007) "A turbulent wake as a tracer of 30,000 years of Mira's mass loss history," *Nature,* 448: 780.

WOH G64, the largest known star: Levesque, E et al. (2009) "The physical properties of the red supergiant WOH G64: The largest star known?," *The Astronomical Journal,* 137: 4744. Searches on the web will list various other stars proposed to be larger. Some brief comments on these other contenders are as follows:

» VY Canis Majoris: listed in *Guinness World Records 2011* as the largest known star, and sometimes suggested to be larger than the orbit of Saturn. However, recent measurements show it to be only 70% of this size. For example, see Choi, Y et al. (2008) "Distance to VY Canis Majoris with VERA," *Publications of the Astronomical Society of Japan*, 60: 1007.

» VV Cephei: similarly has been claimed by some sources to be the largest star, but new observations show its diameter to be about two-thirds that of WOH G64. See Hagen Bauer, W et al. (2008) "Spatial extension in the ultraviolet spectrum of VV Cephei," *The Astronomical Journal*, 136: 1312.

» V838 Monocerotis: has been measured to be about the same size as WOH G64, but the measurement was extremely uncertain and imprecise, partly because the distance to this star is not really known. See Lane, B et al. (2005) "Interferometric observations of V838 Monocerotis," *The Astrophysical Journal*, 622: L137.

IC 1011, the largest known galaxy: Uson, J et al. (1990) "The central galaxy in Abell 2029—An old supergiant," *Science*, 250: 539.

The Sloan Great Wall, the largest known structure in the Universe: Gott, J et al. (2005) "A map of the Universe," *The Astrophysical Journal*, 624: 463.

5. Extremes of Speed

WASP-12b, a fast-orbiting planet: Hebb, L et al. (2009) "WASP-12b: The hottest transiting extrasolar planet yet discovered," *The Astrophysical Journal*, 693: 1920.

I calculated the velocities of this and all other currently known exoplanets using the catalog at exoplanets.org, accessed October 30, 2011.

Orbit of the Sun around the center of the Milky Way: Reid, M et al. (2009) "Trigonometric parallaxes of massive star-forming regions. VI. Galactic structure, fundamental parameters, and noncircular motions," *The Astrophysical Journal*, 700: 137.

SDSS J090745.0+024507, the fastest hypervelocity star: Brown, W et al. (2005) "Discovery of an unbound hypervelocity star in the Milky Way halo," *The Astrophysical Journal*, 622: L33.

RX J0822.0-4300, the fastest known neutron star: Winkler, P & Petre, R (2007) "Direct measurement of neutron star recoil in the oxygen-rich supernova remnant Puppis A," *The Astrophysical Journal*, 670: 635.

The Oh-My-God Particle, the highest-speed cosmic ray: Bird, D et al. (1995) "Detection of a cosmic ray with measured energy well beyond the expected spectral cutoff due to cosmic microwave radiation," *The Astrophysical Journal*, 441: 144.

6. Extremes of Mass

GJ 1245C the lightest known star: Henry, T et al. (1999) "The optical mass-luminosity relation at the end of the main sequence," *The Astrophysical Journal*, 512: 864. I caution that this is a difficult question to answer definitively, and that there are other candidates for this title. However, based on the available evidence and the quality of the data, I judged GJ 1245C to be the most likely holder of this record.

A1, the heaviest known star in the Milky Way: Schnurr, O et al. (2008) "The very massive binary NGC 3603-A1," *Monthly Notices of the Royal Astronomical Society*, 389: L38. Heavier stars have been claimed, but their masses have been determined indirectly or approximately. A1 is the heaviest star with a precise, reliable mass measurement (which usually requires a star to be part of a binary system).

WR 102ka, the star in the Milky Way with the largest initial mass: Barniske, A et al. (2008) "Two extremely luminous WN stars in the Galactic center with circumstellar emission from dust and gas," *Astronomy & Astrophysics*, 971: 984.

S5 0014+813, a supermassive black hole: Ghisellini, G et al. (2009) "The blazar S5 0014+813: A real or apparent monster?," *Monthly Notices of the Royal Astronomical Society*, 399: L24. Accurately measuring the masses of supermassive black holes is very difficult, so it's hard to be completely definitive as to the heaviest black hole, but S5 0014+813 seems to be the black hole with the largest, most reliable measurement.

Abell 2163, the heaviest known galaxy cluster: Holz, D & Perlmutter, S (2010) "The most massive objects in the Universe," arXiv.org/abs/1004.5349.

7. Extremes of Sound

Abell 426, source of the deepest note in the Universe: Fabian, A et al. (2003) "A deep Chandra observation of the Perseus cluster: Shocks and ripples," *Monthly Notices of the Royal Astronomical Society*, 344: L43.

The first sounds in the Universe: all the calculations are my own, but I acknowledge three sources of information that I found especially useful:

» General cosmology: Rich, James (2010) *Fundamentals of Cosmology* (2nd ed.), Springer, Berlin.

» Cramer, John "The sound of the Big Bang," http://faculty .washington.edu/jcramer/BBSound.html, accessed May 5, 2010.

» Whittle, Mark, "Big Bang acoustics," www.astro.virginia .edu/~dmw8f/BBA_web/index_frames.html, accessed May 5, 2010.

8. Extremes of Electricity and Magnetism

YZ Canis Minoris, a spectacular flare star powered by magnetism: Kowalski, A et al. (2010) "A white light megaflare on the dM4.5e Star YZ CMi," *The Astrophysical Journal*, 714: L98.

HD 215441, the most magnetic Ap star known: Babcock, H W (1960) "The 34-kilogauss magnetic field of HD 215441," *The Astrophysical Journal*, 132: 521.

PG 1031+234, the most magnetic white dwarf: Latter, W et al. (1987) "The rotationally modulated Zeeman spectrum at nearly 10^9 gauss of the white dwarf PG 1031+234," *The Astrophysical Journal*, 320: 308. It's hard to be absolutely definitive, because of the difficulty of these measurements; a review of the situation is provided in Jordan, S (2009) "Magnetic fields in white dwarfs and their direct progenitors," *IAU Symposium*, 259: 369.

SGR 1806-20, the most magnetic magnetar: Kouveliotou, C et al. (1998) "An X-ray pulsar with a superstrong magnetic field in the soft gamma-ray repeater SGR 1806-20," *Nature*, 393: 235.

The giant flare from SGR 1806-20 in December 2004: Gaensler, B et al. (2005) "An expanding radio nebula produced by a giant flare from the magnetar SGR 1806-20," *Nature*, 434: 1104.

PSR J0537-6910, the highest-voltage neutron star: my own calculation, using the data in Marshall, F et al. (1998) "Discovery of an ultrafast X-ray pulsar in the supernova remnant N157B," *The Astrophysical Journal*, 499: L179.

The high voltage around supermassive black holes: Straumann, N (2008) "Energy extraction from black holes," *AIP Conference Proceedings*, 977: 75.

Extragalactic jets: Kronberg, P et al. (2011) "Measurement of the electric current in a kpc-scale jet," *The Astrophysical Journal*, 741: L5.

9. Extremes of Gravity

GRO J0422+32, the black hole with the strongest gravity: Gelino, F & Harrison, T (2003) "GRO J0422+32: The lowest mass black hole?," *The Astrophysical Journal*, 599: 1254. There have been claims made of the discovery of an even smaller black hole, XTE J1650-500 (see www.nasa.gov/centers/goddard/news/topstory/2008/smallest_blackhole.html). However, the discoverers of this result later revised their calculation and retracted their claim—see Shaposhnikov, S & Titarchuk, L (2009) "Determination of black hole masses in galactic black hole binaries using scaling of spectral and variability characteristics," *The Astrophysical Journal*, 699: 453.

The weak gravity between the Milky Way and Andromeda galaxies: Van der Marel, R & Guhathakurta, P (2008) "M31 transverse velocity and local group mass from satellite kinematics," *The Astrophysical Journal*, 678: 187.

The future fate of the Milky Way and Andromeda galaxies: Cox, T & Loeb, A (2008) "The collision between the Milky Way and Andromeda," *Monthly Notices of the Royal Astronomical Society*, 386: 461.

The weakest known orbit, between SDSS J113342.7+482005 and SDSS J113403.9+482837: my own calculation, performed using the data presented in Karachentsev, I & Makarov, D (2008) "Binary galaxies in the local supercluster and its neighborhood," *Astrophysical Bulletin*, 63: 299.

10. Extremes of Density

Crystalline structure of white dwarfs: Metcalfe, T et al. (2004) "Testing white dwarf crystallization theory with asteroseismology of the massive pulsating DA Star BPM 37093," *The Astrophysical Journal*, 605: L133.

Long chains of atoms on a neutron star's surface: Salpeter, E (1988) "Hydrogen in strong magnetic fields in neutron star surfaces," *Journal of Physics: Condensed Matter*, 10: 11285.

Extreme strength of a neutron star's crust: Horowitz, C & Kadau, K (2009) "Breaking strain of neutron star crust and gravitational waves," *Physical Review Letters*, 102, id 191102.

Nuclear pasta in the interiors of neutron stars: a good overview is given in Lamb, F (1991) "Neutron stars and black holes," in D Lambert (ed.) *Frontiers of Stellar Evolution*, 20: 299 (published by the Astronomical Society of the Pacific).

Cosmic voids, sites of the lowest densities in the Universe: Hoyle, F & Vogeley, M (2004) "Voids in the two-degree field galaxy redshift survey," *The Astrophysical Journal*, 607: 751.

INDEX

ABOUT THE AUTHOR

Bryan Gaensler is an award-winning astronomer and passionate science communicator, who is internationally recognized for his groundbreaking work on dying stars, interstellar magnets, and cosmic explosions. A former Young Australian of the Year, NASA Hubble Fellow, and Harvard professor, Gaensler is now an Australian Laureate Fellow at the University of Sydney, and is also Director of the Centre of Excellence for All-sky Astrophysics. He gave the 2001 Australia Day Address, was named one of Sydney's 100 most influential people for 2010, and in 2011 was awarded Australia's Pawsey Medal for outstanding research by a physicist under age 40.

www.extremecosmos.net